PERSPECTIVES ON
CLINICAL
TEACHING

The author:
Dorothy W. Smith, R.N., M.A., Ed.D.
Professor of Nursing, and Chairman,
Department of Medical-Surgical Nursing
Rutgers, the State University
of New Jersey

PERSPECTIVES ON
CLINICAL
TEACHING

By DOROTHY W. SMITH

SPRINGER SCIENCE+BUSINESS MEDIA, LLC

ISBN 978-3-662-38714-6   ISBN 978-3-662-39596-7 (eBook)
DOI 10.1007/978-3-662-39596-7

Copyright © 1968 Springer Science+Business Media New York
Originally published by Springer Publishing Company, Inc. in 1968
Softcover reprint of the hardcover 1st edition 1968

Library of Congress Catalog Card Number: 68-21142

To the Memory of
CLAUDIA D. GIPS

*She would rather light a candle
than curse the darkness*

Preface

This book is intended for those whose work involves clinical teaching. Its primary orientation is toward instruction in undergraduate nursing programs. However, because of the increasing number of auxiliary personnel who give direct patient care, and therefore the growing importance of their clinical instruction, one chapter discusses adaptations of clinical instruction that are necessary when teaching this group.

Generally, when referring to the nurse, I have used the pronoun "she" simply because the profession is still preponderantly made up of women. But because of the considerable proportion of men among non-professional workers, the pronoun "he" is used in the one chapter just mentioned.

As programs for preparation of nurses multiply, and as the responsibility for clinical instruction is increasingly assumed by faculty, a growing number of clinical teachers are needed, and more opportunities exist for them to adapt clinical instruction more fully to educational purposes. Through this book I am trying to share my experiences and views on some aspects of clinical instruction with other teachers, particularly with those who are beginning to teach in the clinical area. Surely, the students' clinical experience has often been misused to provide service for hospitals. But I believe—and the belief is reflected in the following pages—that this clinical experience, when used to carry out educational goals, can truly become the heart of the nursing curriculum. I believe it deserves the efforts of our most skillful teachers.

Since the book is not based on a systematic review of the literature and since I am not discussing the literature, there is no formal bibliography at the end of each chapter, but instead a short list of suggestions for further reading. However, the ideas presented derive from many sources, including literature in general education and nursing, and I have tried to point out the ways in which I have adapted and used some of this material in my clinical teaching. It would be impossible, however, for me to list or credit the sources that have, over the years, influenced my thinking. For example, it is not possible for me to say which book, article or lecture led me to fresh understanding of a particular aspect of clinical teaching.

I have excluded some aspects of clinical teaching because they are extensively dealt with in other publications; I have omitted others because I have not yet defined my own philosophy regarding them. Naturally, I have emphasized those aspects of clinical teaching that are of greatest interest to me, and with which I am most familiar. I hope the book will stimulate other clinical teachers to present different dimensions of the subject, and to take issue with those I have presented, thereby enlarging the opportunities of all clinical teachers to learn from each other.

I wish to express my appreciation to those who helped in the preparation of this book.

Miss Gertrud B. Ujhely, Director of the Graduate Program in Psychiatric Nursing at Adelphi University, gave unstinting encouragement and offered many useful suggestions. Her ideas on nursing and teaching have influenced mine over the years of our professional association, and some of her influence is reflected in this book.

Mr. Curtis Freed, who at the time was a senior in the baccalaureate program in nursing at Rutgers University, shared his views on the particular challenges facing men students of nursing. Mrs. Harriet Weinssen, Instructor of Practical Nursing at Morris Hills Regional School in New Jersey, gave valuable assistance with the chapter on teaching the nonprofessional student. Mr. Bernhard J. Springer of the Springer Publishing Company showed interest and confidence in the project, and unusual patience in awaiting its completion. Others from whom I received valuable assistance cannot be named individually. They are my students, my patients, and the many colleagues who have shared their ideas with me over the years.

DOROTHY W. SMITH

Roselle Park, New Jersey

January, 1968

Contents

Part I

THE CHANGING CLINICAL
ENVIRONMENT

1

Introduction

The importance of clinical teaching

Clinical teaching is the most vital part of nursing instruction. It is during work with patients that the student develops her understanding of relevant theory and begins to recognize the problems and rewards inherent in nursing practice. During clinical work, particularly, the student becomes aware of her reactions to patients and ways in which these reactions affect the care she gives. During her clinical practice the essential unity of nursing can become clear to the student through a variety of nursing situations. In each new setting, and with each new category of patients, her attention is focused at first on the differences; for example, the difference between bathing an infant and bathing an adult. But as the student moves from one clinical area to another, she discovers the unity or common core of functioning with patients and families which characterizes nursing. Promoting comfort, assisting with various therapies, helping the patient to understand what is happening to him and how he can help himself, and sustaining him during his experience—these are all aspects of nursing wherever it is practiced.

During her clinical work the student comes in contact with disturbing situations: the patient who has been deserted by his family; the child with leukemia; the lack of trained personnel and supplies to meet requirements for nursing care; and the callousness which sometimes develops in those constantly in touch with human misery.

Some challenges in clinical teaching

Teaching students in settings where practice is far from ideal, and helping students to deal with such difficult problems as the care of a terminally ill child, have always been part of clinical teaching. In addition, today's clinical instructor is expected to adapt to changing theories of nursing and nursing education, and to different approaches to clinical instruction. Instructors are often asked to teach in ways which are radically different from the way they were taught, and to question many traditional beliefs and practices; for instance, it was once considered sufficient to send the student to various services for a specified time in the belief that working with patients and staff would impart to the student the necessary knowledge and skill. Now the teacher is expected to identify the particular clinical learning experiences necessary to meet certain educational objectives. At one time, considerable emphasis was placed upon contact with many patients with a wide variety of diagnoses. Now greater concern is expressed about providing opportunity for more extended and comprehensive experience with a particular patient.

One pressing problem facing nursing educators today is that of determining how much clinical practice is necessary. Although there is general consensus among educators that the amount of practice formerly provided was educationally unnecessary, they have by no means agreed upon the degree of proficiency in nursing practice which should be possessed by the graduate, nor upon the amount of clinical practice necessary to achieve such proficiency. One fact is certain: The amount of time devoted to clinical practice in undergraduate programs has been decreasing steadily. It is therefore especially important to consider carefully the techniques of clinical instruction, so that the time spent on this part of the student's education may be used as effectively as possible.

Changes involving the method and content of clinical teaching often create difficulties for the instructor. By its very familiarity, the clinical situation calls forth responses which have become almost second nature to the teacher, who may respond automatically on the basis of her previous experience as a student and staff nurse. For instance, when the head nurse asks, "Will the students relieve each other for lunch, so that their patients are not left unattended?" the teacher may answer "Yes" before remembering that her students are beginners, and that giving them alternate lunch periods will mean that some students will be on the ward unsupervised while she is at lunch. The request to "cover the lunch period," which at first seemed reasonable, is shown upon reflection to be in conflict with students' need for instruction and patients' requirement for supervision of beginners who care for them.

Regardless of how similar or how different her own program was from that of her students, the teacher occasionally finds herself comparing the performance of students with her own, as she remembers it. "Could I ever have been that slow?" "Wasn't I more responsible than that?" Such questions may come to mind as one observes a student who is slow or hesitant, or one who seems to need more reminders than her classmates. We evaluate our students by the standards we hold as teachers, not according to those we held as students. In addition, we may have had more familiarity with the hospital than students have today. A student who spends a great deal of time at the hospital quickly becomes acquainted with the routines, and with the location of equipment. Such familiarity can lead to a quick, sure demeanor which is often lacking in a beginner who spends only 15 hours a week at the hospital. The teacher may occasionally wonder, as she listens to students express some of their feelings about patients, "Did we react so emotionally to things when we were students?" Perhaps the difference lies not so much in the students' reactions as in the fact that they are now encouraged to discuss them more freely, and that teachers work more closely with students clinically, and therefore have more opportunity to observe emotional responses.

One of the most difficult challenges of clinical teaching involves helping students to recognize the value of their work with patients. Our society accords highest esteem to those health professionals whose work dramatically brings about cure. The surgeon who repairs a defect in a child's heart, the research worker who discovers a new antibiotic which will save many lives, the psychiatrist whose skill enables a hospitalized patient to return to work—all are justifiably esteemed for the usefulness of their work. Many students who enter nursing do so in order to share in the drama of restoring others to health, but they have not yet differentiated between the nurse's role and that of other health workers. It takes time, experience, and the guidance of skillful teachers to help the student realize that although nurses share in this drama, their work emphasizes the daily care which enables patients to help themselves more effectively, rather than through dramatic intervention leading to sudden cures. Of course, nurses also have opportunities for spectacularly altering a person's prognosis, or saving his life; for example, stopping a hemorrhage, restoring a heart beat, or suctioning away a mucous plug. However, for most nurses, most of the time, the miracles are quiet, and are often known only to the nurse and the patient. No newspaper headlines announce, for instance, that a public health nurse's patience and encouragement over many months enabled an elderly man who was afraid of surgery to consent to a cataract extraction, yet what that nurse achieved was just as essential in restoring the man's sight as the surgeon's intervention. In addition to measures which help patients to make use of resources available for care, are those which provide support during an ordeal. While most of the laurels may go to a drug, or to some complicated surgery, many a patient has said to the nurse, "But I don't know how I could have gone through it without you." Often the gratitude is unspoken, since many patients are shy and awkward about expressing it, or are too sick to express it at the time the nurse is caring for them.

One fallacy which has been difficult to overcome is that, because many nursing measures seem relatively undramatic, they require little skill. The beginning student often shares this belief. Few people would say, "Anyone can do it," when observing the implantation of a pacemaker, but many may think, when observing a nurse sitting beside a child in the emergency room, "Anyone can do that." However, the sensitivity and skill of that nurse are not automatic; they are the result of careful observation of the child and skill in reassuring him. The beginning student may be all agog to observe a cardiac catheterization or to learn to use the monitoring devices in the coronary care unit, but her face may fall when the teacher suggests that she instruct an old lady in how to care for a chronic leg ulcer, or sit with a child who is awaiting surgery.

As she works with students, the clinical teacher gradually learns to help them see the possibilities for helping others which lie within their own role, and the value of this care to the patient. She can, for example, show by her own care of a patient how he can be assisted to gradually learn to do his own dressing. By her comments to students concerning the care they give she can help them recognize the changes in a patient's condition which their care has brought about: the patient who was afraid to cough following chest surgery but who is now bringing up the mucus which could have caused atelectasis; the mother who brings her child to clinic because the student's sensible suggestions enabled her to find someone to care for her other children on the day of the clinic appointment; and so on Unless measures are taken to help students recognize the value of what they can do within the scope of nursing, some of them do not develop their skills as fully as they might, but instead, look toward the roles of others to obtain interest and satisfaction from their work. They become the nurses who, for example, peer through the speculum to see the cervical erosion being pointed out by the gynecologist, rather than holding the patient's hand and encouraging her to breathe deeply and relax during the pelvic examination.

One difficult aspect of clinical teaching involves helping students deal with situations in a way which may run counter to natural human inclination, and to society's values and customs. For example, most people prefer to be around others who are cheerful, attractive, entertaining, independent, and self-controlled. We tend to avoid those who are despondent, helpless, anxious, disfigured or malodorous. Nevertheless, the student must learn to care not only for the joyous mother and her adorable new baby, for example, but also for patients who have had illegal abortions, are incontinent or confused, or are radically disfigured by such diseases as cancer.

Compassion is the bridge which can carry the nurse's skill and concern to those who need her most. But compassion is not enough, nor is skill enough to enable the nurse to handle such situations, particularly when she must work with them over a considerable period of time. Stamina, a quality which is sometimes not sufficiently stressed, is also necessary. Helping students to develop the compassion and stamina needed to enable them to use their knowledge and skill with patients who, by society's standards, are unappealing, is a challenge faced by the clinical instructor. Demonstrating not only her skill, but also her compassion and stamina—by example when she herself works with such patients—and encouraging students to discuss their reactions to these difficult situations are measures which can help the teacher meet this challenge.

The clinical teacher must be both practitioner and teacher—a duality of roles which requires a disciplined command of theory, facility in its application, and ability to help others develop knowledge and skills. Persons so qualified do not at all resemble the traditional image of a clinical instructor—the relatively inexperienced practitioner, just beginning to develop her teaching skills. The clinical teacher of today and of the future requires assistance, during her graduate study, in learning how to apply concepts to the actual practice of clinical instruction. Guided experiences with various aspects of clinical teaching can save the future teacher much time, and help her to function more effectively in her first teaching position.

One may well question the reasons for lack of emphasis given to clinical teaching. Why is it, for example, that clinical teaching problems often receive scant attention in some faculty discussion groups, and in some graduate programs preparing teachers? Possibly it is a reflection of the fact that comparatively greater esteem is accorded to administration and classroom teaching than to nursing practice. Perhaps it can be traced to our culture, which places higher value on intellectual pursuits than on pursuits involving ministrations with one's hands—unless these ministrations are performed in a way that is dramatic and out-of-the-ordinary. In addition, our nursing tradition has, in some respects, de-emphasized the value of clinical practice while according higher prestige to classroom activities. In the past much of students' clinical practice was repetitive, and its purpose was to provide hospital service. Classroom experiences, on the other hand, were educationally controlled, and it was therefore natural to emphasize their value. Everyone in the clinical setting was supposed to do clinical teaching, and it was not expected that one required special preparation in order to do it. Certain physicians and nurses were selected for classroom teaching; to be selected accorded recognition and prestige. Although some nurses and physicians fulfilled admirably their responsibilities for clinical instruction, others either were not especially interested in teaching, or did not have time for it due to their obligations to patients. In any case, the responsibility for carrying out a program of clinical instruction was often not clearly fixed, whereas the responsibility for conducting certain classes was definitely assumed by the nurses and physicians doing the teaching. In some instances, clinical instruction was viewed as a favor bestowed by busy staff members when time permitted, rather than as an obligation assumed by the school in order to provide necessary instruction for students.

When it became possible for nursing educators to control clinical learning experiences, the result was like harnessing the power of a mighty river. The opportunities to use clinical experiences for students' learning increased enormously, and the need for qualified clinical teachers grew accordingly.

Other forces were also at work to raise the esteem in which clinical practice was held. The profession itself began to show greater respect for it; for example, the American Nurses Association placed increasing emphasis on the development of clinical skills by such measures as providing clinical sessions during state and national convention programs. Graduate education in nursing began to emphasize the development of excellence in a particular field of clinical practice. Such developments showed recognition of the fact that patients benefit not from what the nurse knows, but from the knowledge which she is able to *apply* in caring for them. The greater attention being paid to the quality of care given, rather than to the length of time the nurse spends with the patient, is further evidence of the concern for development of clinical expertise. The well-worn phrase, "The nurse is with the patient the most," has been too often used glibly without sufficient regard for the quality of care given. For example, one patient who had an electric pacemaker remarked, "There are plenty of people milling around looking at this equipment, but a lot of them don't seem to know how it works, and they seldom pay any attention to me—they only look at the machine." This patient did not lack the presence of nurses, but he lacked skilled nursing care. In contrast, a distraught patient spoke with the public health nurse for fifteen minutes on the telephone but, in that time, the nurse conveyed enough support and gave enough useful suggestions that the patient's apprehension was relieved and she was able to handle her problem effectively.

The changing legal status of students and the gradual increase in the number of clinical teachers have affected the responsibility of teachers for the quality of care given by their students. When students exchanged service for education they were legally considered employees of the hospital.[1] When students pay for tuition, board and room, and carry out only the clinical work which is required for their learning, the likelihood of their being considered hospital employees is remote. This change sharpens the responsibility of the teacher, who realizes that neither she nor the student is afforded the protection, in the event of a suit, of being viewed as a hospital employee. When one teacher was available for one hundred students spread throughout the hospital, the responsibility she was expected to assume was quite different from that of the teacher who works with seven students who are all on the same ward. In the latter instance it is natural that the teacher will be held more accountable for the quality of care given by her students.

These changes challenge the teacher to provide supervision which is careful enough to protect patients, but which is supple enough to foster students' development of responsibility for their own actions, as well as their creativity in trying different approaches to patient care. In such a situation, the instructor's role as teacher-practitioner is emphasized, while in the program in which she has many students on many different wards, her role is necessarily confined largely to planning with staff for students' clinical learning experiences, and coordinating plans for their clinical experiences. The teacher who instructs a small number of students is necessarily more involved with the care of assigned patients. Dealing with excoriated skin around a colostomy opening and talking with distraught family members become problems for her to cope with—as well as for the student to learn about. Most of us find the care of certain types of patients more rewarding or challenging than that

[1] Lesnik, M. J., and Anderson, B. E. *Nursing Practice and the Law*. 2nd ed. rev. Philadelphia, J. B. Lippincott, 1962.

of others. It is especially important that the teacher avoid unwitting tailoring of clinical assignments to correspond so closely with her own interests that some aspects of the course receive greater attention clinically than others.

Most students today do not have as much opportunity to feel a part of the ongoing work of the clinical agency as they once had. Although in the past their status as junior staff members often interfered with learning, it also provided opportunities for the development of a great sense of pride in their contribution to the workload. Since experiences such as running a ward alone at night (because there is no one else available) are—fortunately—becoming less common among students, it is important to show recognition of the contributions they *do* make to patients' welfare: the healing of a decubitus ulcer, thanks to a student's meticulous care; the revival of a patient's interest in learning to use his artificial leg, because of a student's concern; the gradual renewal of a depressed patient's interest in his surroundings as the student works with him over a period of several months. Selection of assignments on the basis of students' educational needs, necessary as it is, must not be allowed to obscure the fact that service is given in the course of learning, and that careful, sensitive care can benefit patients.

Currently, emphasis is being placed upon the importance of the teacher's interaction with students as a means of helping them to develop effective relationships with patients. Consequently, the demonstration of such qualities as patience and support in her relationships with students is essential. It may, in fact, be the most important contribution the teacher can make to students' learning, since it cannot be supplied by any text or teaching machine. However, functioning as a role model for students may seem overwhelming to the teacher (particularly if to her the word "model" connotes faultless perfection) since she, too, functions under stresses which sometimes interfere with her ability to demonstrate these qualities. At such times it may help to remember that students have many role models, among whom are their parents and other teachers. This realization, while not denying the importance of consistent demonstration of the characteristics one advocates to students, can help the teacher avoid overestimating the importance of an occasional slip.

Purposes of this book

The purposes of this book are to highlight some of the challenges in clinical teaching, and to consider some ways of dealing with them. The material is oriented toward instruction of undergraduate students of nursing. However, one chapter on the clinical instruction of nonprofessional nursing personnel has been included. The author's viewpoint is that every clinical teaching program, regardless of the category of nursing personnel being prepared, should be geared to the educational needs of the students in order to make maximum use of the available time and resources and thereby to prepare the best qualified nursing personnel possible.

No attempt has been made to include material which does not directly involve clinical teaching, although it is recognized that such matters as formulation of objectives for the total curriculum, and for the courses which comprise it, provide the framework for the development of both classroom and clinical experiences; the reader is referred to other sources for such material. The aim of this book is to share one teacher's views and experiences in dealing with such challenges of clinical instruction as team teaching, planning clinical assignments, conducting conferences, and working with clinical agencies. Suggestions for further reading are included at the end of each chapter.

It is hoped that this book will:

• Stimulate the experienced teacher to further crystallize her own views in the process of responding to the ideas of a colleague.

• Assist the beginning teacher to deal more effectively with some of the daily problems of clinical instruction.

• Provide those who are starting to teach in the newer types of nursing education programs with an orientation to some of the challenges facing the clinical teacher in such settings.

• Convey to those interested in clinical instruction the belief that this aspect of teaching is not only laced with problems, but filled with opportunities for satisfaction in working closely with students and patients.

2

Team Teaching

There is growing recognition of the importance of having well-qualified faculty members to carry out clinical teaching and to relate clinical and classroom learning. This recognition has fostered the practice of team teaching, a system which makes it easier to meet the need for a faculty large enough to handle the clinical instruction of small groups of students. In previous years it was not unusual for one teacher to be responsible for classroom instruction of forty or fifty students; clinical instruction was carried out largely by nursing service personnel. The broadening of content in some nursing courses (such as the development of a course in the care of mothers and children, rather than separate courses in obstetric and pediatric nursing) has also fostered team teaching, because of the necessity for participation by faculty with different areas of clinical preparation. The effort to integrate various concepts throughout the curriculum, such as those dealing with nurse-patient relationships, has furthered the practice of having a group of faculty with different areas of specialization share in clinical and classroom teaching.

Team teaching has affected the role of faculty, and their working relationships with each other and with students. This chapter will consider some of the effects of team teaching on the instructional program. Although these effects are especially pronounced when a group of faculty is responsible for teaching a single course, many of the topics dealt with in this chapter affect the working relationships of faculty, whether or not they share the responsibility for one course with several colleagues.

ADVANTAGES AND DISADVANTAGES OF
TEAM TEACHING

Advantages

Team teaching has many advantages. The sharpening of insight, crystallization of ideas, widening of interest and knowledge, and validation of evaluation of students' performance are important benefits for faculty engaged in group teaching. For students it can mean stimulating exposure to different points of view and different emphases which even the most versatile teacher cannot provide alone. Some teachers are more skilled and interested in the care of elderly, long-term patients; some are more adept at caring for children. Teaching in a group can allow each faculty member to contribute more fully in her area of particular interest and competence. Teachers can give each other support when difficulties arise during the course. For instance, certain experiences may not be yielding the learning which faculty had anticipated, and students and faculty may both be experiencing dissatisfaction. Sharing their views about the problem can help the faculty members clarify what the difficulty is, and what measures may help to alleviate it. Nor is the sharing all in the area of difficulties and problems. Satisfactions gained from teaching a course have an added glow when shared by those who worked together to attain them.

Orientation of new faculty members is facilitated by team teaching. The opportunity to work closely with colleagues helps them learn about the philosophy and purposes of the course and about the particular characteristics of various clinical agencies. Team teaching also helps relatively inexperienced teachers to contribute effectively to instruction, since they work closely with experienced faculty members. Providing continuity of instruction when a faculty member is ill is facilitated by team teaching, because teachers who are oriented to the purposes and learning experiences of the course can continue the program during their colleague's absence.

Team teaching also allows for flexibility in assigning students to work with particular teachers. A student who is extremely timid about relating to those in authority, for example, may learn more effectively from a junior faculty member than from a senior member. A student who seems very competitive toward faculty, and who seeks many special privileges, may be difficult for the beginning teacher to deal with; this student may require the firmness and confidence of a more experienced instructor.

Disadvantages

What are some undesirable consequences of team teaching? The amount of time and energy required for meetings of the group leaves less of both for individual work and study. When many hours are taken up by meetings, a teacher may find herself preparing a lecture hurriedly, and using only the literature readily at hand. If this pattern is repeated frequently, the quality of classroom teaching is bound to suffer. A certain flattening of ideas and diminution of sparkle inevitably set in, as all of us know who have carried a teaching schedule which did not allow sufficient time for thought and study. Another result of the devotion of many hours to team meetings is that individual conferences with students may be curtailed, due to time pressures. For a group to work together successfully, it is usually desirable that there be consensus on major decisions. How much individuality of teaching is lost, and how many interesting ideas are sacrificed in order to reach a consensus? The toll in both these areas is often high, and constitutes a loss both to faculty development and to students' opportunities for learning. In addition, some flexibility of teaching may be lost. Implementing a program of agreed-upon learning experiences can limit the teacher's opportunity to use unexpected clinical events to their fullest potential for student learning.

If the faculty group is large, occasions for each teacher to lecture may be so infrequent that the opportunities to develop skills in classroom teaching are few. Also, when a course is taught by several instructors, it is sometimes difficult for each member to have a sufficiently active part in developing the course. The faculty member who consistently teaches with seven or eight colleagues may miss the satisfaction of assuming the entire responsibility for planning and conducting a course.

Despite these disadvantages, team teaching has proved itself effective in the clinical instruction of nursing students. The challenge is to get it to work smoothly, with benefits to both instructors and students; to accomplish this requires emphasis, throughout the team endeavor, on the essential unity of all nursing.

Emphasizing the Unity of Nursing

Recently, there has been a growing interest in having faculty from different clinical areas teach, as a team, certain nursing courses. This approach seems appropriate and promising when courses are developed with emphasis upon the common aspects of nursing, regardless of setting, rather than upon the differences related to a particular setting. Instructors in public health nursing, psychiatric nursing, and nursing of children, for example, may jointly present a course dealing with growth and development of the normal child, and care of children with physical and mental illnesses in the hospital, other community agencies, and in the home. Even though her clinical teaching may embrace all facets of the course, each teacher's classroom contribution is in the area of her special competence; this brings unique satisfaction to her, as well as wider learning to students.

Such an approach to teaching presents opportunities for

faculty and students to emphasize the unity of nursing, and to apply principles to varied clinical situations. Its use seems particularly applicable to undergraduate programs preparing the nurse to practice as a generalist in nursing. Each faculty member must have a clear idea not only of her particular contribution to the course, but of the way in which this contribution can be blended smoothly into the whole. She must also be able to make applications of her subject in settings with which she may be relatively unfamiliar.

All of us have some real limitations to our versatility, and these limitations make an important difference in the quality of our teaching. Because of the need for greater numbers of faculty in clinical teaching, as contrasted with classroom teaching, there is more likelihood of being asked to give clinical instruction in an area outside one's own field than of being asked to conduct classes on a topic outside one's specialty.

How broad can one expect faculty preparation to be? Does the teacher who chose to work with adults find it difficult to work with children? If so, how effectively can she carry out clinical teaching on a children's ward? If, in a burst of enthusiasm, faculty agree upon teaching a broadly oriented course, without considering the demands which such teaching will make upon them, the purpose of the course may later be eclipsed as each faculty member runs to her own specialty for cover. The concepts and attitudes portrayed to students are then the opposite of those intended and students may conclude that the unity or wholeness of nursing is a myth perpetuated for students, to be shed with the student uniform. On the other hand, when students observe faculty who have a breadth of knowledge about nursing and who exemplify its application in a variety of fields they have an excellent opportunity to gain appreciation for the unity of nursing.

Although the difficulties of this shared teaching are great, so are the advantages. Students can be helped to develop appreciation for the matrix, or ground-substance, of nursing, and also for the highlights which come from looking at one aspect of nursing intensively. Seeing the various facets of a nursing situation illuminated by the specialized knowledge of faculty members from different fields is enlightening to the student, because it emphasizes the interrelationships of varied aspects of the patient's nursing requirements.

Producing a wholeness in the teaching of a course involves having enthusiastic and competent specialists teach each major aspect of it. Otherwise, the peaks of specialized knowledge may be flattened in a pedestrian manner because the course content does not fall in any one particular teacher's area of interest. Also, the integration of content should not result in diminution of the hard core of factual knowledge required for competent practice; for example, pharmacology may be presented as a separate course or as an aspect of all nursing courses but, in either case, students should learn the pharmacologic actions and therapeutic uses of commonly used drugs.

All teachers must keep constantly before them the broad purpose of the course, and make a continued effort to support the entire course as well as their own special contribution to it. Respect for the value of all aspects of the course, and for the competence of those whose area of specialization is different from one's own, is the foundation for success in this kind of teaching; it also enhances the students' growing awareness of the essential unity of nursing.

It is especially important for faculty to have a clear understanding of the breadth of teaching which is expected of them. Those who have never taught the skills of bathing, bed-making, and taking vital signs may be asked to carry out such teaching clinically. Those whose teaching has been oriented primarily toward the beginning skills in physical care may be expected to also provide instruction in nurse-patient relationships. There should be opportunity to broaden one's own knowledge and skills, if necessary, before one begins to teach a new course. This not only helps teachers to participate more effectively and more comfortably, but it also fosters their appreciation of the value of the varied aspects of the course.

Roles of the temporary and permanent team members

Sometimes the temporary participation of faculty from other fields is sought to help broaden the scope of a course. Public health nursing faculty, for instance, may participate in teaching the course in psychiatric nursing in order to provide for the inclusion of more material concerning the home, the family, and community agencies. Faculty should have a clear understanding about whether their joint participation in teaching is considered temporary or permanent. If it is considered to be temporary, the major emphasis should be upon helping the faculty who are primarily responsible for the course to develop it more broadly. If it is expected that representatives of different fields will continue to participate in teaching the courses in future years, the development by each specialist of her contribution, and the meshing of it with the material presented by others, assumes particular significance.

Influence of the dominant values of the school

The dominant values in a particular school also affect team teaching and group decisions concerning relative emphasis on various subjects. When the faculty of a school is strongly oriented toward physical science, for example, the vigorous group support given to this aspect of teaching may result in lessened emphasis on psychosocial content, regardless of the skill or perseverance of the faculty member who presents this material in an integrated course. On the other hand, if the faculty values particularly the psychosocial content of the curriculum, physical care and technical skills, as well as factual knowledge concerning pathophysiology, may be de-emphasized. These subtle influences may not be reflected in course outlines, or in stated objectives of a program. What is actually stressed, for instance, in clinical conferences, can vary greatly depending on the values of the faculty. Over the several years required for undergraduate education, these differences in emphasis can have significant effects upon students' competence and attitudes toward nursing. For example, students may view highly technical aspects of nursing, such as care of patients who have renal dialysis or heart surgery, as being the acme of professional functioning (if this view is held by most of their teachers), but scorn and possibly be less skilled at care of long-term patients. Such attitudes can affect recruitment of students to various fields, attracting many to one sphere of nursing while discouraging them from learning about opportunities in other fields.

The values held by the faculty group determine somewhat the degree to which they feel free to admit lack of competence or lack of interest in an aspect of nursing. In one school, it may be unthinkable for a teacher to state that her interest and competence do not lie in the area of nurse-patient relationships. In another school, all faculty members may respond to persistent pressure to increase their knowledge of physical sciences by methodically taking additional courses in these fields, and by not admitting any lack of zeal for applying this content to nursing.

Determining Time Priorities

Teachers need opportunities to share ideas with colleagues. They also need time away from the group to think, read and plan. Team teaching underscores the need to determine the relative amount of time faculty should spend in individual study and in such group activities as meetings to collaborate in developing objectives, plan course content, coordinate instruction, or evaluate results of teaching. This is a persistent need in all fields of education, not only nursing education. It is not a question of which emphasis is good and which is bad, but of the relative amount of each of these two types of activity which is most likely to lead to effective teaching. The stimulation of ideas which comes from a lively discussion with colleagues is as necessary as quiet times for study and reflection. In today's extroverted society opportunities for quiet reflection are exceedingly rare, whatever one's field of employment. However, despite the pressures exerted by society toward more and more group activities, colleges and universities have continued to emphasize the value of individual study and research. Whatever subject they teach, university faculty are expected to allot considerable time for the individual study and scholarship so necessary to creative teaching, as well as to participate in the group planning essential to coordination of the educational program and the development of the curriculum.

The values held by the faculty determine, to some extent, the priorities which will be placed upon individual and group work. As far as possible, this emphasis should be the result of deliberate choice, with recognition of the gains and losses involved, rather than a tired capitulation to demands for either type of emphasis. The priorities placed upon group and individual endeavor should be in harmony with the values of the institution; for example, faculty in schools of nursing affiliated with colleges and universities must take account of the emphasis which the institution places on individual scholarship.

The dilemma of allocating sufficient time to group and individual work presents a particular challenge to those engaged in nursing education. The nurse-faculty member in a collegiate setting is especially likely to experience heavy (and to some extent, conflicting) demands for individual scholarship and group participation because the requirements of the institution in regard to research, scholarship and publication are added to a teaching load which has already been made arduous by the addition of responsibilities for clinical instruction. Thus, she finds it difficult to allocate time for the group meetings so essential to team teaching and to other aspects of nursing instruction which also stress group effort; for example, meetings between nursing service personnel and teachers for the purpose of planning the students' clinical experience.

Because of these time pressures, it is important for nurse faculty members to consider whether a task requires group activity, or whether it can be handled by an individual teacher. Is it necessary, say, for a group of seven teachers to spend many hours developing a final examination, or could one teacher prepare a tentative set of questions and bring it to the group for their recommendations and approval? Measures which lead to more effective use of time actually spent in meetings are also important. All members should make every possible effort to delineate the group's responsibilities, to define clearly the purposes of each meeting, and to expedite the accomplishment of tasks for which meetings are called. There should be a clear understanding as to whether the responsibility for such action as the decision to change an aspect of a course rests with the faculty group teaching the course, or whether they must seek approval of the change from a curriculum committee. Ordinarily, decisions concerning the over-all purpose of a course, and the subject matter included, are the province of the faculty as a whole, or of a faculty committee. Decisions about how these purposes will be implemented are usually made by the faculty teaching the course.

Effect on the teaching load

How does team teaching affect teaching load? The decrease in the number of lectures for which teachers are responsible is somewhat offset by the necessity for them to orient themselves to the content of all lectures so that they can plan their own classes to avoid duplication and gaps in content, and help students apply classroom learning to their clinical practice. How teachers achieve this detailed knowledge of the lecture content of a course varies. Some audit all classes given in the course; others obtain the information by conferences with their colleagues, and review of lecture notes. Auditing lectures is the most time-consuming method, but it is also an effective way to learn about each teacher's orientation and emphasis, as well as the students' questions and responses to the lecture. The inexperienced teacher who has never taught the course before can benefit from auditing her co-workers' lectures, while the experienced teacher who has taught the same course previously will probably find that discussion of content in group planning sessions and informal conferences with her colleagues suffice, thus conserving time which is needed for other aspects of her work.

Correlating Clinical and Classroom Instruction

In addition to considering allotment of time for various tasks, the teaching team determines how and when the group of students will be divided for clinical and classroom work. These decisions must be guided by the amount of student participation required for certain learning experiences, the provision of opportunities for faculty to contribute in their areas of special skill and interest, and the necessity for continuity of student-faculty contact, particularly during clinical practice. It is preferable, for example, for the entire group to attend a lecture given by the faculty member best prepared in the subject than for four different teachers to present the same lecture to small groups of students. This is feasible, however, only when the clinical facilities are adequate to provide practice with the type of situations being discussed in class. To achieve this correlation, it may be necessary for some classes to be presented to small groups of students at the time when clinical experience is available to them. Such classes as those dealing with the care of premature infants may have to be presented to small groups, because of the limited number of students who can be accommodated for clinical practice at any one time. In some situations, the faculty may choose to divide the class into sections for all lectures as well as for clinical practice and conferences. While this facilitates correlation of classroom and clinical teaching, it may result in inefficient use of faculty time, and poor coordination of lecture content. Because of the increased number and variety of lecture topics for which faculty are responsible, the quality of classroom teaching may suffer, particularly if each faculty member is also responsible for a great deal of clinical teaching.

When considerable student participation is necessary for effective learning, dividing the class into small groups is essential. Clinical practice, and clinical conferences in which each student presents data, are examples of situations which necessitate working with small groups. However, some aspects of clinical teaching can be presented to larger groups of students by such methods as televised demonstrations of the teacher giving care to a patient. The latter method not only accommodates more students, but also conserves faculty time, allows all students in the group to observe the teacher who is most skilled in the particular aspect of care being demonstrated, and avoids the tedium sometimes experienced by faculty when a demonstration must be repeated many times.

Regardless of the way in which a class may be divided for clinical and classroom experiences, all teachers must collaborate in developing the course objectives and requirements, descriptions of necessary learning experiences, and criteria for evaluation of students' achievement. If students are divided into separate sections for all classroom as well as clinical work, the need for collaboration among faculty is intensified since the lack of opportunities for sharing in the actual teaching can lead to marked differences in course requirements and in criteria for evaluation of performance from one class section to another.

The need for communication among faculty engaged in team teaching appears to be limitless. Even after a series of meetings during which faculty share, and agree upon, their expectations concerning student performance at different stages of the course, students may state that different criteria are being used by each faculty member. Students in one section of a course may consistently have papers returned to them which are not written in acceptable English, while in another section this aspect of evaluating written work may receive scant attention. Despite efforts to develop agreed-upon criteria for evaluation, the ironing out of disagreements and differences of opinion may take years, rather than the few months allotted to a particular course. If the same group teaches the same course repeatedly, greater opportunity exists for the sharing of ideas which can lead to well-coordinated teaching.

Factors Affecting Team Relationships

When a group of faculty collaborate in teaching a course, the working relationships among them and the way the group is organized are important. A particularly sticky question involves who is in charge of the course, and what being in charge means. Does it mean that the person in charge develops the entire teaching plan, and then orients others to it, down to the last detail? This may be an appropriate approach for a professor who has several teaching assistants, but would hardly be suitable when others in the group are full-time faculty members. In the latter instance it would seem more feasible for one person to coordinate the activities of the group, such as initiating meetings with hospital personnel, and for all faculty to participate in planning the course.

The role of the junior faculty

Perhaps the crucial question in connection with achieving smooth team relationships concerns the role of the faculty, particularly junior faculty members. Should they be expected to spend several years auditing lectures and conducting clinical laboratories, or should they begin by contributing to the development of the courses they are teaching? Which approach is most likely to help them develop the qualities expected of the experienced teacher?

If these qualities include independent thought, and the ability to assist in the development of courses which comprise both classroom and clinical work, it would seem most useful for them to begin promptly to contribute ideas concerning the courses they are teaching, and to take an active part, with guidance by more experienced faculty members, in classroom as well as clinical teaching.

Changes in nursing education have fostered eagerness among faculty who have graduated from the newer programs to participate actively in planning, teaching, and evaluating the courses with which they are involved. Greater emphasis is now placed upon the importance of stating one's views and thinking independently than was the case in nursing schools in the past. The student who has been encouraged to think independently and to speak up in group discussion does not usually, upon becoming a faculty member, become reluctant to state her views in a faculty meeting. Eyebrows may be raised that one so new is also so audacious.

However, continuity of the teaching program must be maintained, and over-all planning for a course must precede the actual teaching. Therefore, in most instances, new faculty members should expect to function within the framework of plans already made, and to suggest major changes only after they have had experience teaching the course, and have acquired knowledge concerning the way in which its content relates to the total curriculum.

Rivalry for popularity with students

The close working relationships inherent in team teaching sometimes gives rise to problems connected with the teacher's quest for popularity with students. While this may not manifest itself in such overt actions as currying favor by lowering standards or raising grades, it has many subtle effects on teaching and learning. Encouraging students to express their criticism of other teachers, and the development of an exaggeratedly social, chum-like manner with students are examples of ways in which teachers may seek popularity. The need for recognition is universal, and popularity with students is one form of recognition.

Manifestations of this problem can be seen in many everyday working relationships. There is the teacher who characteristically springs upon her colleagues some tidbit which students have told her—for instance, that another teacher's assignments are boring—with an air which implies, "The students tell me things they don't tell you." Such a person can create an explosive situation when another teacher works with the same group of students, and each derogates the competence of the other. Then there is the group leader who makes vague references to student dissatisfaction with a course, but who is unwilling to state specifically what the problem is, thus leaving her colleagues with no data to use in remedying the situation, but with doubts about the effectiveness of their teaching. Such problems are minimized when mutual respect and frank communication exist among faculty members, and when problems arising in the teaching process are dealt with in a manner which helps the teacher to increase her understanding and her skill.

What are the effects of these rivalries on students? They may find themselves drawn into disputes which can seriously impede their learning. After a bit of gossip with a faculty member about another teacher, they may find it hard to face the latter at the next day's class. Students should be encouraged to express their views in a responsible manner directly to the faculty member involved, or to the person to whom the faculty member is immediately responsible, or both; it is up to the students to decide whether the matter warrants such action. Criticizing others is a maneuver sometimes used to deflect attention from one's own poor performance, and some students seem to use criticism of their teachers for this purpose.

Limits of group decision-making

How far can decision-making by the faculty group

extend without encroaching on prerogatives which are properly those of an individual faculty member? The answer to this question is basically the same whether it is considered in relation to team teaching or in relation to faculty working relationships as a whole. There is general agreement among nurse educators that the entire faculty should decide on the purposes of the program and design the curriculum in light of these purposes, and that each faculty member (or each group of faculty members) should plan specific courses in accordance with this over-all framework. This concept provides ground rules for a continuous interplay between decision-making by the entire faculty, and by individuals responsible for each course. If either the individuals or the faculty group fail to assume sufficient responsibility for making decisions which are appropriately theirs, the program suffers. Faculty committees should not make detailed plans for carrying out the objectives for specific courses. One can envision the frustration of a faculty member who has specialized in a particular field of nursing, on finding that many details of lecture and clinical content are decided by a committee composed of faculty from various fields, thus seriously interfering with her opportunity to contribute in her area of special competence.

Effect of Team Teaching on the Clinical Setting

How may team teaching affect the clinical setting in which students practice? Both patients and staff of the agency can benefit from contacts with faculty who have varying points of view and who share their knowledge of how to handle various nursing problems. When the teacher of public health nursing, for example, has contact with students and staff on a pediatric ward, she may stimulate their efforts toward coordination of care from hospital to home. Sharing of knowledge and different viewpoints can occur in many ways, such as attendance by the staff at clinical conferences about patients, and by the daily work with patients which is carried out by students and faculty. A broadening of skills and perceptions of nursing which can improve the quality of nursing care may gradually occur as staff, students, and faculty work together.

The clinical instructor who teaches a course outside her usual field faces certain difficulties, however. The customs and nursing practices in the new environment may be quite different from those with which she is familiar and, unless she orients herself to these differences, she may unwittingly confuse students and impair working relationships with staff. Some of these differences concern matters which are of relatively minor importance in terms of patient care, but which can affect working relationships with staff. For example, it may not be unusual for the nurse to join the patient in smoking a cigarette while she talks with him in a psychiatric setting, but if she smoked while talking with a patient on a medical-surgical unit, the staff might view this action as a breach of professional etiquette.

While such matters seem almost too trivial to warrant a faculty member's attention, they have important effects upon relationships with staff, as well as upon the teacher's ability to guide students in behavior acceptable in different settings. The situation is analagous to observing the customs of one's hosts when in their home, even though one may not adopt these customs in one's own home. The faculty member who functions in different clinical environments is in a good position to help students examine various customs of staff and various ways of working with patients which are peculiar to certain settings, to consider some of the reasons for the differences, and to differentiate between those which are merely customary in a certain setting, and those which are based on the requirements of patients. One custom which varies with the circumstances is the wearing of street clothes while caring for patients; this may be perfectly appropriate when one is caring for long-term patients, but it would not be appropriate when caring for those in an intensive care unit because of the possibility of the spread of infection, as well as the necessity for protecting street clothes from soiling when working with patients who require a great deal of physical care.

These differences extend from customs of dress and etiquette to differences in the role of the nurse in working with patients and other health professionals. For instance, it may be appropriate for the public health nurse who has worked with a cancer patient and his family and physician over a period of many months to assume an active role in helping the family decide what to tell the patient about his condition at various stages of his illness. In contrast, when a nurse works with a patient who has been diagnosed as having cancer in the hospital setting, and is not well acquainted with him, his family or his physician, her assumption of such an active role concerning what the patient should be told could interfere with a coordinated plan of care, and impair her working relationship with the patient, his family, and the physician. Unless such differences are explored and clarified, students may be confused when it appears that faculty from various fields are making conflicting suggestions for handling clinical situations. Teachers who share the responsibility for a course can assist each other to recognize the varying approaches necessary in different clinical settings, and in this way provide more smoothly coordinated learning experiences for students.

Conclusion

Repeated sensitizing of each member of the faculty to the needs of the patient in his particular situation is one safeguard in preventing over- or under-emphasis on certain aspects of nursing during team teaching. If the patient is in shock, it is important to check his vital signs frequently; this does not rule out the need to comfort his family, and later, when his physical condition has improved, to help the patient understand what happened to him. It seems reasonable to expect the professional nurse to encompass these varied aspects of care, and to view them as worthy of her time and talent. Another safeguard involves knowing ourselves, and our abilities, well enough to be able and willing to say what we can and cannot do. Frankness is essential, and one way of fostering it is to recognize and accept each person's particular abilities. One person may be a good generalist; another may not be able to stretch her thoughts and her efforts to an area outside her own. Perhaps the challenge lies in helping each faculty member do what she can do best, and arranging the teaching program in a way which will allow each person to make her maximum contribution. Whatever the theoretical concept of curriculum development is, it can succeed only to the extent that it accords with the values and abilities of the faculty who implement it.

SUGGESTED READING

Bair, Medill, and Woodward, Richard G. *Team Teaching in Action.* Boston, Houghton Mifflin Company, 1964.

Geitgey, Doris A. "Some Thoughts on Team Teaching in Nursing Education," *Nursing Outlook 15*:66, October, 1967.

Heller, Melvin P., and King, Imogene. "Team Teaching," *Nursing Outlook 13*:50, October, 1965.

Herge, Henry C. *The College Teacher.* New York, Center for Applied Research in Education, 1965.

Johnson, Edgar N. *et al. Freedom and the University.* Ithaca, N. Y., Cornell University Press, 1950.

Lyons, Veronica. "The Junior Instructor," *Nursing Outlook 12*:33, June, 1964.

Nuckolls, Katherine B. "Tenderness and Technique Via T.V.," *American Journal of Nursing 66*:2690, December, 1966.

Redman, Barbara K. "Conflicts in Clinical Teaching in Nursing," *Nursing Forum 4*:48, No. 2, 1965.

Reinkemeyer, Sister Mary Hubert. "An Inherited Pathology," *Nursing Outlook 15*:51, November, 1967.

Schoenberg, Bernard. "Consultation in a Multidisciplinary Group Teaching Program," *Nursing Forum 5*:65, No. 4, 1966.

Whyte, William H. *The Organization Man.* New York, Doubleday, 1957.

3

Teacher-Student Relationships

The relationship between student and teacher should be one which fosters learning and the provision of skilled nursing care. Clinical teaching adds another dimension to this relationship in that the student and teacher become, in many ways, junior and senior partners in caring for patients. Both satisfactions and responsibilities are involved in this relationship. There is, for example, the rewarding opportunity for the teacher to share the student's fresh insights, questions, and enthusiasm. The teacher must also, however, accept responsibility for accurately assessing the student's requirements for instruction and supervision, and for dealing with embarrassing or sometimes hazardous situations resulting from the student's uneven performance and inexperience. The instructor shares in the results of the student's performance, sometimes with pride and sometimes with disappointment. A test of this relationship occurs when the student's actions lead to an undesired outcome. The teacher must then assist the student to do what is necessary to rectify the consequences of her action and to prevent recurrence of similar incidents. This must be done in a way which provides support for the student without condoning her mistake, or de-emphasizing its seriousness, and without undermining the student's confidence in her own ability. The teacher must convey to the staff her concern and her sense of responsibility about the incident, without implying a lack of confidence in the student's over-all ability.

Successful teacher-student relationships are the sine qua non for effective clinical teaching. The objective of both participants is the development of the junior partner into a competent nurse with judgment and skill. But, within the process, both partners are limited, and it is the responsibility of each to recognize and cope with both sets of limitations.

Overcoming Some of the Limitations of the Instructor

Dealing with ones' own fallibility

Just as important, or perhaps more so, than the concepts taught verbally, is the teacher's actual functioning with patients, students, and staff. The clinical situation, with its pressing requirements for patient care, makes it difficult for either teacher or student to take refuge in verbalisms and the kind of theorizing which defies implementation. Human functioning is richer and more varied than any textbook description, but it also encompasses failure to measure up to the standard of ideal performance. Admitting that one does not know a certain fact or how to work a particular piece of equipment requires judgment concerning its possible effect upon the student. For instance, the first time a student bathes a patient is probably not the best time to tell her one does not know how to operate the patient's electric bed. (Such points should be investigated ahead of time or, if this is not possible, it should be done away from the student, who is probably at this particular moment looking to the teacher as a panicky swimmer looks to a lifeguard.) On the other hand, as the student gradually becomes accustomed to clinical work, it is essential to let her in on the fact that one does not always know the necessary points, and then seek the information together. The teacher's work is usually not so limited to one ward or even to one hospital, nor is her competence so broad that she is spared the frequent necessity of asking questions and looking up information. To do this quite openly can help students avoid developing the unrealistic expectation that they will know, or should know, everything once they have graduated.

The teacher's imperfect functioning is not confined to occasional lack of factual knowledge or technical know-how. There are times when the pressures of work lead an instructor to be less patient and perceptive than the situation requires. When such a reaction of impatience is obvious to both student and teacher, it is better to make a brief apology than to pretend that the event has not occurred. Admitting one's shortcomings is more likely to gain the student's respect, and it may help her to acknowledge occasions when her own actions do not measure up to the requirements of the situation. It can also lead to a more relaxed relationship between teacher and student, because both acknowledge their human fallibility.

The subject matter nurse-teachers deal with calls attention to the gap between the human qualities they advocate and the extent to which most of them are able to demonstrate these qualities in their relationships with students and patients. Because the subject matter presented by the clinical teacher is so closely related to such personal attributes as acceptance and thoughtfulness, students sometimes develop unrealistic expectations concerning the extent to which their clinical instructors can demonstrate these qualities. In short, the clinical teacher is faced (to a greater extent perhaps than faculty in some other fields) with the problem of how to deal with her "feet of clay." Trying to conceal them, or being embarrassed about them, seems only to make them more obvious, and more in the way during her teaching. The only alternative, then, seems to be to acknowledge one's clay feet when they intrude upon one's teaching, without consistently pointing to them or dwelling upon their existence.

This challenge is made more insistent by the fact that most undergraduate students are in late adolescence, a time of life when young people seek association with adults who exemplify the values they profess. Some students have, in the past, had associations with adults who conveyed, "Don't do as I do; do as I say," and thus they entered nursing already sensitized to the divergence between the values adults advocate, and those which they practice.

Students should be encouraged to discuss the problems they may encounter in dealing with the discrepancy between what teachers and members of the staff do and what they advocate. They may, on occasion, indicate in various ways that they are disappointed in the way the teacher or staff member is handling a situation. These are the times to encourage students to discuss the matter of learning from imperfect role models, since this is something they will have to face throughout their lives. If such discussions occur periodically when the occasions arise, students can be helped to avoid accumulating resentment over the discrepancy between what they expect of faculty, and the way faculty measure up to these expectations. Gradually, students can be helped to develop more realistic expectations, and greater tolerance for human frailty—their own, as well as that of others.

Focusing on one's primary role

The focus of the student-teacher relationship should be on teaching and learning. The teacher stays with and listens to the student who becomes upset over a clinical incident, and assists the student with aspects of learning which are difficult for her; but helping her to deal with personal problems which may be affecting her learning is ordinarily the function of the counselor. Precisely where the limits should be in a specific situation will be determined by the teacher in relation to her competence, the time she has, the way she perceives her role, and the availability of counseling services. It is essential to recognize the limitations of one's time and energy, and the demands of the teaching situation. Lack of focus upon one's primary role can lead to a gradual erosion of the quality of teaching, thus diluting the particular contribution one is best prepared to make to each student's growth.

Sometimes faculty attempt, unwittingly and unrealistically, to combine the roles of teacher and counselor. However, these two roles have significant differences. The teacher must, of course, accept responsibility for the results which the student achieves clinically—how well she communicates with a patient, how effectively she gives an injection, and so on. The teacher must also require a certain discipline, impose certain time limits for achievement, and evaluate the student's performance. These responsibilities weigh especially heavy when dealing with a student who does not perform adequately and who therefore receives a low grade, or fails the course. The counselor, on the other hand, is not responsible for setting standards for performance, or for evaluation. The teacher's relationship with students is enhanced by honesty concerning her role. In other words, she should not imply that her role involves merely helping students to achieve at their own rate, and to the best of their abilities, when actually it also involves setting time limits, establishing minimum standards of performance, and giving a grade. Unless these aspects of the teacher's role are clarified with students at the beginning of the course, sometimes it is not until much later, when a student receives a low grade, that she takes full account of this aspect of the teacher's role which was in operation all along, and which should have been made explicit at the start of their relationship.

Assuming these responsibilities, however, does not prevent the teacher from showing interest and concern for the student as a person. Sometimes without realizing it (although the teacher may spend many hours helping an individual student), the interaction is always task-centered; the emphasis may always be, "How well have you mastered that clinical problem we spoke of last week?" rather than a relaxed and interested, "How are things going?" which allows the student to voice some of her concerns about the program, and which shows the teacher's interest in her as an individual. Sometimes, when students who are viewed by the faculty as having received a great deal of individual attention comment, "But the faculty aren't interested in us," the divergence in views can be traced to such intense focus on the task that the student feels a lack of interest in her as a person, regardless of how much time a teacher may have spent with her.

Helping the student to trust the imperfect model

The teacher who recognizes her own fallibility recognizes, in the bargain, that she is an imperfect model for her students. This recognition should be more than humbling—it should also make her a better teacher by compelling her to examine how she can help students to trust her despite her limitations. One way to achieve this crucial relationship of trust is to "tune in" to the way students are handling their clinical assignments.

It is essential to convey to students that direct assistance and observation of their performance form an important part of clinical teaching, and that this teaching will be done straightforwardly and without apology. A surreptitious or apologetic approach to supervision can make both teacher and student acutely uncomfortable. Seeing the teacher's shoes on the other side of the curtain and realizing that she is listening to a conversation with a patient can be an unnerving experience for the student, unless the teacher has already planned with the student that she will listen, unannounced, to conversations with patients in order to help the student develop skill in talking with them.

In contrast, the teacher may hear a student's conversation with the head nurse while reading a chart at the nurses' station. In this instance, the teacher is making no apology for or effort to conceal her presence, and the student can take account of her presence even though not addressing her. Such openness of supervision can help the student develop trust. This is an important consideration, because the student is being observed and assisted with activities with which she is inexperienced and perhaps clumsy.

The beginning student's awkwardness and clumsiness

affect the development of trust in her relationship with the teacher. The nursing student's performance is on display— a fact which can cause anxiety, and sometimes humiliation and anger. Most students have many embarrassing experiences in the process of learning to practice nursing. These blunders may be obvious to many persons including the teacher, the physician, the head nurse and the patient. One student, who was assisting the physician with a dressing for the first time, noticed that he held out his hand. Not realizing that this gesture indicated a request for a gauze pad, the student shook his hand! Imagine the feelings of another student whose hand was visibly shaking as she prepared to give an injection, when her patient inquired, "Are you going to rivet it in?" Every nurse has her private dossier of embarrassing experiences, many of which were acquired during her student days.

As the years pass, the acute embarrassment and humiliation which one feels on these occasions lessens, and the teacher may regard these experiences of her students as humorous or ordinary and may, therefore, be startled at the intensity of some students' reactions to them. Of course, many students are able to take such events in stride, but it is not unusual for others to feel severe "loss of face" and, as a consequence, to become anxious, angry, and temporarily more clumsy. The teacher can help by trying to anticipate the requirements of situations and briefing students ahead of time, so that the number of embarrassing incidents can be kept to a minimum, and by offering suggestions tactfully and privately, to prevent them from becoming an additional source of embarrassment.

Sometimes the teacher is surprised, too, at the way a student responds to suggestions, even when they are made tactfully and privately. After all, from the teacher's standpoint the criticism is clearly for the student's benefit, and it seems only reasonable to expect that it will be received graciously. Some students, however, react angrily, and may show even greater ineptitude and awkwardness for a short time because of their anger. Sometimes the student is angry with the teacher for making the suggestions. Often, though, she is angry and impatient about her shortcomings, and feels humiliated because she has not fulfilled her own expectations. In such instances, the teacher can help by pointing out that a certain degree of awkwardness goes with every learning process, and that the clinical practice situation lends itself to a display of one's clumsiness to patients and staff.

Some students have highly unrealistic expectations in regard to their own performance. For instance, they may envision themselves performing their first catheterization with the smooth dexterity shown on a film they have just watched, and be surprised when the teacher suggests taking along a few extra catheters, "just in case." Sometimes these difficulties are made worse by demonstrating a procedure in an unrealistically ideal environment very different from the one which the student will encounter on the ward. While such a demonstration away from patients has its usefulness in permitting discussion and questions as it is carried out, it does not provide the necessary realism and should, therefore, be supplemented by a demonstration with a patient, using the standard ward equipment. As the student observes the teacher making the adaptations required by the situation and by the patient's response, she may develop a more realistic appreciation of the difficulties and awkwardness inherent in some nursing situations and, consequently, may approach these procedures with more reasonable expectations concerning her own performance.

Other ways in which the teacher may help the student develop trust include being consistent in what is expected of her and showing respect for her. The student will know more surely where she stands if lapses in screening a patient are consistently called to her attention, rather than being commented on one day and disregarded the next. In a ward where patients and staff are not called by their first names, it is inappropriate to use the student's first name merely because she is young. (Calling the student "Miss" also helps the teacher deal with situations in which students ask the teacher if they may call her by her first name. A matter-of-fact reply, with possibly a touch of humor, usually clarifies that a friendly student-teacher relationship is different from that which exists between pals.)

Tuning in to the individual student

It is important for the teacher to keep her antennae receptive to the way each student is working. Such tuning in results from attentiveness to the way the student is handling her clinical assignment, from avoidance of unnecessary interference with what the student is doing, and from minimizing distractions which deflect one's attention from the student's performance.

Becoming too active (when such activity is not useful to patient or student) can so alter the situation that the opportunity to see how the student will proceed is lost. Consider, for instance, the student who is about to assist a postoperative patient to turn on his side so that she can wash his back. Although this is a procedure which both the student and patient know how to manage, the teacher instructs the patient in how to turn, and assists him to do so. Such intervention—sometimes resorted to in order to help the teacher feel more comfortable and useful—may unwittingly imply to the student that she is not proceeding capably, and it prevents the teacher from observing how the student would have carried out the procedure. Too much and too frequent intervention can also result in directing the student's energies away from her patient in order to respond to the flow of instructions, warnings, and reminders from the teacher. Far from being passive, though, the teacher must be actively alert to how the student is proceeding, and to whether and in what ways she requires assistance.

Students must become able to function in a situation in which the teacher comes and goes, but is available to them when they seek help with patient care. Especially when they are new to clinical work, students need to be told of the purpose of the teacher's periodic presence when they are giving patient care, what she is observing, and how she will use her observations in helping them to learn. Individual conferences, repeated at intervals throughout the course, help the student get the "feel" of the observations the teacher is making, and of how these can be useful to the student in improving her nursing practice. Observations and suggestions are, of course, also shared with the student by means of frequent brief discussions during the clinical laboratory. When suggestions and criticisms are offered promptly during practice periods, the student is helped to realize that the teacher is not accumulating examples of mistakes to confront her with at a later time. This confidence fosters the student's willingness to deal with unfamiliar aspects of patient care.

Recognizing one's limitations in working with
certain types of students

Most teachers find that they work with some types of students more effectively than with others. Factors affecting one's ease in working with various kinds of students include cultural background, personality, and the values one holds. A teacher who is a member of a particular cultural group, for example, may find it comparatively easy to understand and respond helpfully to some of the problems experienced by a student who is also a member of that group.

The personalities of the student and teacher also influence their interrelations. Sometimes the teacher works more productively with students whose reactions are similar to her own, but the opposite may also be true. A shy teacher may work very effectively with shy, timid students, and find it difficult to work with aggressive students. A teacher who has had more than the average amount of difficulty in acquiring manual dexterity may be particularly patient and understanding with a student who also has this difficulty. However, another teacher who has experienced this problem may be reminded too acutely of her own earlier shortcomings when she teaches clumsy students, and be especially severe in her criticism of them.

Values are important, too, in affecting a teacher's ability to work productively with certain types of students; for instance, a teacher who is deeply committed to her career may find it easier and more satisfying to teach students who express an interest in attending graduate school, than those who plan to retire from active practice upon graduation in order to devote full time to home responsibilities.

Recognizing one's limitations, considering individual differences when planning clinical assignment of particular students to particular teachers, and realizing that students whom one finds difficult to teach have opportunity to work with other teachers who do not share this difficulty, are helpful in coping with this problem. It is inevitable that teachers will work more effectively with some students than others, but recognizing the factors responsible for the difficulties can sometimes help the teacher to deal with them. If, for example, the teacher who tends to be impatient with students who are slow and awkward is aware of this same tendency in herself, she is more likely to recognize that although a student may lack manual dexterity she may be particularly skillful in some other area such as answering patients' questions. When the clinical instruction is shared with one or more colleagues it is sometimes possible to arrange students' clinical assignments so that the various teachers work with the types of students with whom they function best. In the instances when this is not feasible it is helpful to remind oneself that those students with whom one was not particularly effective may be the very ones with whom the next instructor will work most productively.

HELPING STUDENTS OVERCOME SOME OF THEIR LIMITATIONS

Channeling reactions to difficult nursing situations

Helping students deal with anxiety-producing clinical situations. Recognizing manifestations of anxiety during clinical practice, and realizing which clinical situations are likely to be upsetting for a particular student or group of students is an important part of the teacher's task. This aspect of her work may be facilitated by sensitive observation of clinical performance and frequent individual conferences with students, during which they are encouraged to discuss their reactions to clinical work.

As students approach clinical practice it is helpful for them to know that the teacher realizes that most people experience some anxiety when approaching a new experience. One teacher noted a great deal of giggling and restlessness at the start of a laboratory period in which students were to practice giving injections to one another. With a twinkle in her eye, she said, "It's that first jab that's the hardest, and most people are nervous about making it." The laughter which followed was out of proportion to the humor of the remark, but it served to "ground" some of the anxiety which had seemed to crackle around the room like electricity. Her comment called students' attention to the fact that most people are anxious in certain situations, and that this is an expected reaction. When a student is assigned individually to stressful situations rather than sharing them with classmates, it is sometimes hard for her to realize that she is not the only one to experience this reaction. She may, for example, approach her first scrub in the operating room literally in a cold sweat, but assume that she is the only one who is responding in this fashion. During group discussion prior to such assignments it is often possible to encourage students to express some of their apprehension about the coming experience. Then, even though they are assigned individually, they are supported by the realization that other students are also nervous about this experience, and that this reaction is quite natural.

It is important for the teacher to be able to differentiate between the degree of anxiety to be expected in the situation, and that which seems out of proportion, and for which the student should be referred for counseling. It is expected that a student may tremble the first few times she administers injections, or that she may be nervous on the first few occasions when she works with a patient in isolation. Particular situations may be distressing to certain students; for instance, a student who ordinarily plans and organizes her work well may appear to be very disorganized in the way she carries out care of a paralyzed patient. She may run back and forth in a never-quite-successful attempt to assemble the necessary equipment, and spend little time with the patient. Calling the student's attention to this, and inquiring about the difficulty can help the student to recognize that she is anxious (if this is the case), and to consider what it is about this patient that makes it difficult for her to be with him. Discussing the difficulty, and then helping the student to plan how she will organize and carry out the rest of his care will probably suffice to enable her to handle the situation satisfactorily.

The fact that a student occasionally loses self-control does not necessarily mean that she is in serious emotional difficulty. The traditional emphasis in nursing upon the "stiff upper lip," has sometimes resulted in unnecessary alarm and dire predictions about a student's suitability for nursing if, on one or two occasions, she has burst into tears or become disproportionately angry. Such an outburst from a student who ordinarily handles herself well may merely indicate that stress is piling up. A combination of staying up late to study for an important exam, having a fight with her boyfriend, on top of a case of dysmenorrhea may cause a usually poised student to lose her aplomb. Talking with the student, once she has calmed down, can help the teacher realize the effect of accumulated stress in bringing about the outburst, and help her avoid over-estimating the importance of the incident.

While it is important not to overestimate the significance of an occasional "off-day," it is also essential for the clinical instructor to be alert to signs that a student is consistently experiencing an unusual degree of anxiety, and to refer the student for assistance with her emotional problems. (See pp. 173-174 for discussion of the teacher's role in referral.) Some of these signs include persistent interference with her ability to apply knowledge to clinical practice, repeated difficulty in organizing her clinical assignments, frequent problems in predicting the results of her actions, inability to modify her behavior in accordance with changed circumstances, difficulties in performance and relationships which repeatedly show themselves in a variety of clinical situations, failure to profit from the teacher's suggestions, and low tolerance for stress. Of course, these kinds of behavior do not necessarily indicate anxiety. The student who does not apply her knowledge to clinical work may not be motivated to help others, although she has an intellectual interest in the subject of nursing. One who has difficulty organizing her clinical work may never have had to carry out tasks at home, and therefore lacks rudimentary skill in organizing and completing a series of tasks involved in patient care. Following are three examples of behavior which may indicate anxiety in the student in the clinical setting.

1. *Interference with ability to apply knowledge.* A student who had had considerable experience with giving medicines, applying aseptic technique, and setting up intravenous infusions was unexpectedly faced with the necessity to add an antihistamine to a bottle of solution to be used for intravenous infusion. The instructor said, "Get the medication ready, and then wait for me to come back before you add it to the infusion bottle." When she returned the teacher found that the student had selected an oral preparation of the antihistamine, rather than one for parenteral use, and was planning to dissolve the tablet in the intravenous solution. The student was shocked when her error was pointed out. It was not necessary for the teacher to discuss at length the fact that the tablet was unsterile, that it might not dissolve thoroughly, and so on, because the student already possessed that knowledge. What was necessary, though, was that the student be helped to pierce through the cloud of

anxiety which momentarily prevented her from using the knowledge she had. Her comment later concerning this incident was, "I was so upset about the patient, and I wanted so badly to do everything right that for a minute there I just wasn't thinking."

2. *Difficulty in predicting the results of her actions.* A student rushed past a sign saying, AUTHORIZED PERSONNEL ONLY, into the operating room suite. She was dumbfounded when the operating room supervisor crossly reprimanded her. The student's comment was, "I don't know why she was angry. I just wanted to see the O.R." In light of her knowledge about transmission of infection, and of the sign which the student said she saw, she should have been able to predict the supervisor's anger at her disregard for hospital rules.

3. *Inability to modify behavior in accordance with changed circumstances.* A student had been caring for a coronary patient for several days. Each day she bathed him, took his vital signs, administered his medicines, and made his bed. One day during her morning briefing, the head nurse said, "Mr. Low is to be fed his breakfast, have his vital signs taken, and be given his medicines. That is all. He had a recurrence of chest pain last night, and a bath might tire him and bring on the pain again." A short time later, the teacher noted that the student was filling Mr. Low's bath basin. The teacher asked, "What do you remember about the head nurse's report on Mr. Low this morning?" The student repeated the head nurse's comments practically verbatim, and only then did it occur to her that, although she had heard the explicit instructions and the reason why a change in the care plan was necessary, she was proceeding with care exactly as she had performed it every other day.

Some students are more likely than others to call attention to their reactions to anxiety-producing situations, and to seek and receive assistance from teachers and counselors. One who does not show the usual indications of anxiety in such situations may well be an example of the student whose need for assistance often goes unrecognized. Nursing tradition has placed such a high value on control of emotions, maintenance of a cheery demeanor and an air of confidence, that the student who seems utterly unperturbed in situations which are ordinarily anxiety-producing, who adopts a breezy, stereotyped manner toward all her patients, and who seems overconfident of her abilities may be viewed as doing very satisfactory work. In contrast, one who is often timid in approaching new situations, and who occasionally overreacts emotionally to clinical problems calls attention to her difficulties.

The latter may seek and receive considerably more assistance from teachers and counselors than the former. Often it takes a great deal of the teacher's time and patience to help the seemingly over-confident student become willing to admit that she too requires assistance and support in learning to deal with patient-care problems. Such a student can be extremely irritating to the teacher. Not only does she tend to tackle situations which lie beyond her abilities without appearing to recognize the implications but, instead of seeking instruction, she may seem to dare the teacher to offer it. While her breezy "Hello-how-are-you" approach may at first give the impression that she relates easily to patients, her relationships with patients tend to remain superficial.

Unless the teacher carries out considerable direct, thoughtful observation of such a student's work it is difficult to provide the kind of concrete suggestions which can help the student to improve her practice. It is easy to acquiesce to her apparent confidence, and although feeling vaguely uneasy about her work, to spend more time with students who seek assistance and receive it graciously. The seemingly over-confident student may respond to the query, "How are you getting along?" with a resounding "Fine, thank you," which is likely to bar any further discussion. It is helpful to plan such a student's assignment in a way that will facilitate unobtrusive observation, because direct supervision often makes this type of student particularly apprehensive. Assigning her to work in the same room with one or two other students, rather than in a private room, will allow the instructor to observe her work while supervising that of the other students, and can provide concrete examples for discussion. It is particularly important to make suggestions to such a student gradually and gently, and to be certain that they are firmly based on direct observation. The student is not likely to be helped by the teacher's comment that she is too superficial in her approach to patients, but she may profit from discussion of a specific example of superficiality in her relationships with patients. One such student, for example, replied to a patient who expressed discouragement over the prospect of a second operation, "Oh, I'm sure you'll be just fine." During discussion of this incident with the teacher, the student could consider what other responses might have been more helpful to her patient.

There are certain additional measures that can be used in helping students learn to deal with anxiety-producing situations:

Encourage the student to admit (if only to herself) that certain experiences are difficult for her to handle. This is one way of helping her begin to cope with difficult situations. The subtle suggestion that "the good nurse" finds every patient care situation rewarding and satisfying can make it difficult for students to admit that they find some experiences distressing. The teacher who conveys recognition that nursing experiences can be disturbing, and that the nurse who is distressed by certain situations is no less worthy, can help students recognize and acknowledge experiences that are difficult for them to handle.

Embarrassment over performing such aspects of nursing as toileting and giving perineal care can be particularly acute among adolescents, but this can be minimized if the student has a thorough knowledge of the procedure and familiarity with the equipment. During a class on toileting, for example, both the bedpan and the urinal should be demonstrated. One student, whose teacher related the topic only to women patients, was asked by a patient to empty his urinal, which was covered by a bedpan cover. Since she had never seen a urinal the student was not sure where the handle was, and she was too embarrassed to remove the cover and look. The result was that she handled the covered urinal so awkwardly that she spilled the urine on the floor.

Let the student observe the way the teacher gives intimate care to a patient. This will help the student learn how to proceed without causing needless embarrassment for the patient or herself. A student who was caring for a patient who was menstruating delayed giving her perineal care, although the teacher had previously discussed with the student this aspect of the patient's care. The teacher, noting the delay and speculating that it might be due to embarrassment, suggested that the student observe her giving the care. The teacher's relaxed but matter-of-fact approach to the patient, her care in draping and screening, and the sure touch which she used accomplished more in indicating how the nurse can minimize embarrassment than any amount of explanation could have done.

Provide opportunity for students to voice some of their concerns about performing certain procedures at the time of the initial demonstration. Sometimes, despite the fact that the teacher recognizes the desirability of discussing students' emotional reactions to such procedures as toileting, allocation of time is handled in such a manner that only the factual and technical aspects of this care are presented at the beginning of the nursing program. Discussion of students' reactions to giving such care may take place a year or more later, perhaps in a course on nurse-patient relationships. Students need to be given a chance to consider their reactions to such procedures before they can be expected to carry them out with sensitivity to the patient's needs.

Certain clinical situations are recognized as upsetting to most people; care of a patient at the time of death is an example. Many of these responses are highly individual, however. One student may be overwhelmed by the antics of a group of convalescent children, but functions well when caring for individual children who are confined to bed. Her classmate, on the other hand, seems to feel helpless when assigned to care of a sick youngster, but enjoys assignment to the playroom. Because of the variability of reactions it is important for the teacher not to decide what experiences should be distressing to students (on the basis of her own responses), but to remain as alert as possible to students' reactions.

Students may find that some clinical experiences arouse worry over their own health. Most of them are in late adolescence, a period of life usually accompanied by considerable concern over body functioning. Some patients' symptoms are not unlike symptoms the student herself may have experienced; for example, caring for a patient with menorrhagia due to fibroid tumors may remind her that she experienced a somewhat heavier flow than usual during her last menstrual period, and cause her to wonder whether she, too, might have fibroid tumors. Sometimes the student mentions such worries to the teacher, who can then suggest that the student consult her physician. Concern about her own health is also affected by her increased knowledge of the causes, treatment, and prognosis of various illnesses. This knowledge may be both reassuring and disquieting: reassuring in the sense that the student is more familiar with terminology, use of various therapies and of what may be expected from them; disquieting in the sense that she becomes increasingly aware of the limits of medical knowledge and skill, and of the fact that not all patients receive optimum care in the light of available knowledge.

When a student is to be assigned to a type of situation which has caused her difficulty in the past, it is usually wise to orient her to the assignment ahead of time with particular care and with ample opportunity for discussion. There are times, however, when discussion serves merely to delay tackling a task. This delay can heighten the student's apprehension. For instance, once one has talked with a student concerning her reluctance to administer injections, has demonstrated and redemonstrated the technique, and has had the student practice hydrating oranges, it may be preferable to have her proceed promptly to give the dreaded injection when the time for its administration arrives. Once the medicine has been injected, discussion about it will not be accompanied by the student's mounting dread of actually pushing the needle through the skin.

Encourage the student to express her reactions to patients. Ideas about what is appropriate for the student to say concerning her experiences with patients have changed, particularly with regard to what may be said away from the patient. Is it acceptable for her to say she felt angry with a patient, or repulsed by his appearance? Is saying that she finds a patient repulsive the only way of expressing this feeling? Is allowing her to make such a comment likely to lead to expression of her feelings to the patient which could be harmful to him? Probably the student who is not encouraged to talk about her reactions to her patients will be more likely to show them in ways which can be harmful to him as well as to her. Talking about her reactions can help her become aware of them, and of the ways she may be expressing them to the patient. By encouraging students to talk with their patients we are increasing their chances for becoming involved in upsetting situations. Discussing their reactions to patients is a necessary and important part of students' learning. Self-control is not antithetical to the expression of some of one's feelings, but has to do with where, when, and to whom they are expressed.

If, when the teacher was a student, she was not encouraged to express her reactions to patients, she may find that although she tells her students it is all right to do so, she is outraged when they express any negative reaction to a patient. She may rush into a discussion of why the patient may be the way he is, without giving the student a chance to finish what she was saying about her own reactions. When this occurs, it is helpful for the teacher to reconsider her own views about this kind of discussion, and perhaps also to talk about the incident with a colleague.

Show your own emotional reactions when this is appropriate. What about the teacher's expression of emotion? Should she be tightly controlled at all times, never showing impatience or frustration? Such an ideal (if indeed it is an appropriate one) is impossible for most people to attain; the real need is an increasing awareness of one's own reactions and of the manner in which they are displayed. Never to show sorrow at a patient's death, or irritation with a seemingly senseless rule can make the teacher seem not fully human. One way of helping students to respond warmly to patients is through contact with people who show emotion, but who usually do so with consideration of the effect upon others. Contact with teachers and staff members who show emotion but know how to channel it and use it appropriately in serving patients, can help students deal with one of their persistent concerns—how to avoid becoming "hardened" despite their contact with human suffering.

Although clinical assignments can be modified temporarily because of a student's emotional problems, each student must demonstrate her competence at the level and with the diversity of experiences necessary for satisfactory completion of the course. If she is not ready to cope with these experiences it may be necessary for her to drop the course, at least temporarily, while counseling continues. The student who has not mastered one part of her program is usually ill-equipped to handle the next part, and continuing to place her in clinical situations may, depending on the nature of the difficulty, be a disservice to patients, as well as to the student.

Handling student complaints

Listening to students complain seems to go with teaching as inevitably as the grading of examination papers. One may be startled that, despite more instruction, carefully supervised practice, and even curtailment of night and evening duty, students continue to complain about other things, such as the keenness of competition in the chemistry course, or the rigors of traveling to clinical agencies. It is especially difficult for teachers whose students have many more opportunities for learning and recreation than they themselves had, to recognize that the complaints are just as meaningful to these students as their own were to them. Reminding students that they should be thankful not to be on 12-hour duty with half a day off on Sundays somehow seems irrelevant to present-day students. The achievements of one generation become the routine expectations of the next. Moreover, many of the changes that have occurred in nursing education do not make the program easier; they merely make the difficulties somewhat different. In some programs, students are not scheduled for clinical work on weekends, but they may spend many a Saturday and Sunday writing term papers. Nurse educators must continue to improve programs of study, but the expectation that students will at some point no longer complain is unrealistic.

What usefulness does griping have for the students? The nursing student is, in many ways, "low man on the totem pole" in clinical situations. Her opinions carry little weight; her actions and decisions are carefully supervised. Griping may be one means of helping her tolerate the little humiliations which come to the novice in any field. The tiny wounds to pride suffered during a clinical day—not remembering how to work the oxygen apparatus; knocking over a drainage bottle; forgetting to fasten a call bell, after being reminded of it only yesterday—can be soothed somewhat by discussing incidents which point out that even experienced hospital staff are not immune to lapses in the quality of care they give, or that equipment is missing or unsatisfactory. (Few hospitals lack for such examples, which very properly require improvement.) The veracity of the examples brought up by students is not the sole consideration. It is also necessary to consider the possible usefulness of the griping in helping students deal with their relative power-lessness and awkwardness in nursing situations.

Realizing that griping has some usefulness can help the teacher listen to it without feeling that she must personally and promptly remedy every problem which students cite; it can also help her avoid thinking that, because students complain, they are necessarily unappreciative of her efforts to teach them. The teacher should listen to the complaints, make an effort to understand them within the context of the students' situation rather than in comparison with her own, and assist students to take whatever steps seem feasible to improve matters. Although it is important for the teacher to listen to students' complaints, to support the truth of their comments when they are true, to bring in relevant factors which may have escaped the students' notice ("The staff nurse had 12 patients; you had two"), and to do what lies within her power to improve the situation, it is also essential for her to see that conferences are not devoted mainly to this kind of discussion, and that adequate time is allowed for dealing with other topics. Lengthy discussions of the failings of others is one way for students to avoid consider-ing what they can do to improve their own present and future practice. In addition, it is hard for students to appre-ciate that there really will be a time when their opinions about nursing will carry more weight, and that they really will have more authority by which to influence care given to patients. Reminding them of this, and of the necessity for learning as much as possible during their student days so that their own practice can be skillful, can help them move on to other topics.

The teacher must make clear that irresponsible com-ments, and those made merely to belittle others will not go unchallenged. An important difference exists between using undesirable situations for discussion in order to understand why they happen and how they may be prevented, and making vague sweeping statements which derogate others. Instead of appearing ignorant of undesirable practices which occur on the ward, or of joining in with expressions of her own frustration over some clinical incident, the instructor should direct the discussion toward the learning which students can derive from the incidents they cite, and toward encouraging acknowledgment of such feelings as anger or disillusionment concerning the situation being discussed.

Helping students adapt their behavior to the clinical setting

Placing limits on students' behavior in the clinical setting is another area of the student-teacher relationship in which some traditional values are being questioned. One may cite, for example, the teacher who was so irked, during her own student days, by the emphasis on clean shoelaces and punctuality that she now goes to the opposite extreme in disregarding personal untidiness and tardiness among her students. Later, these concerns may absorb more of her time and attention, rather than less, because students' appearance and lateness begin to irritate the nursing staff, and may even lead them to lose some confidence in the teaching program. To emphasize such qualities as personal neatness and courtesy sufficiently to help students develop these habits without stressing them so much that attention is diverted from other important matters, requires judgment and an ability to adapt these requirements to changes in practices among nurses in general. At one time it was unthinkable for a nurse to smoke while in uniform. But how reasonable is it for today's teacher to insist on enforcing this rule among her students, when many staff nurses smoke in the hospital dining room, and in staff lounges? Although she may not agree with this practice, the teacher who tries to enforce a "No smoking in uniform" rule will probably succeed only in arousing resentment among her students at the difference in the standards held for them and for graduates. Rather than making rules about student conduct which are markedly different from those adhered to by most practicing nurses, it is preferable for nurse educators to work with other nurses in education and service in order to deal more effectively and more fairly with the problem of setting standards.

Helping students develop confidence,
responsibility, and individuality

Particular problems of the collegiate setting. The wisdom of nursing instructors is taxed by the necessity to safeguard the welfare of patients by providing adequate supervision, and to do this in a way which offers expanding opportunities for the student to use her own judgment. This challenge is especially difficult to meet in collegiate programs where faculty and students are often involved in proving the worth of the program to a skeptical clinical community. In such a situation, even the smallest error can seem like a catastrophe; the strain experienced by students and faculty can be intense, and can itself predispose to errors. The faculty may find themselves caught between the criticism of staff that students are being babied instead of taught, and the attempt to see that every detail of assigned care is carried out in a way which is above reproach.

These problems can be handled more smoothly if faculty and agency staff are not misled by the notion that collegiate students are, or should be, free of the inconsistencies of adolescents, or less prone to human error than students in other nursing programs. Perhaps it is natural to develop some unrealistic expectations of students in a relatively new type of program which has been launched with great difficulty, expenditure of energy, and emotional investment. As time passes, expectations may become more realistic. One faculty member commented at the end of her first semester of teaching in a collegiate setting, "Well, they really are kids after all. I guess I was expecting them to behave like graduate students at the age of eighteen."

The gradual acceptance of the program, and the increased confidence of faculty and students which results when several classes of students pass licensing examinations and practice successfully, can ease the strain, and can enable faculty to consider more fully the problem of helping students gradually to become more self-directing. Decisions concerning the point at which the student should be allowed to proceed without direct supervision are necessary; for instance, if the student has been observed giving four intramuscular injections correctly, is it necessary to observe her administration of the fifth? Differentiating the procedures which, if improperly performed, can lead to harm, from those which cannot, is important. Neatly covering the patient with an oxygen tent but forgetting to turn on the oxygen is in a different category from folding a bath blanket untidily. What is hovering in one instance may be wise surveillance in the other.

Occasionally, in her efforts to make certain that patient care is carried out satisfactorily, the teacher may lose sight of the fact that most nursing students are highly motivated to give good nursing care. Most students can, if encouraged to do so, take on some of the "checking up" themselves; they can be expected to review their own charting for completeness before leaving the ward, for example.

One aspect of helping the student acquire independence involves assisting her to develop her own way of working with patients, within the framework of accepted practices. As she begins her nursing practice the student must imitate, but gradually she should develop a way of caring for patients which is uniquely her own. To help the student achieve her own style of nursing, it is essential that she learn to distinguish between those things which must be done in a certain manner and no other, and those which can be handled somewhat differently by different nurses and still produce a favorable result. One student's playful exuberance may be just as effective in encouraging an anorexic youngster to eat as her classmate's soft-spoken gentleness; each student may be communicating her concern and interest in the child in a manner to which he can respond. It is easy to criticize a student for doing something in a way which is different from the manner in which one would have proceeded. Before deciding that the student is in error (except in situations where her error is obvious and requires prompt correction), it is important to ask her why she is proceeding in this manner, and to consider whether her approach may be acceptable, even though different from one's own.

Working with any patient is a highly individual process, involving not only the technical aspects of care, but also the relationship between student and patient. Too often in the past teachers have attempted to provide models (often determined largely by the patient's diagnosis) for students to follow in caring for patients. For instance, the teacher might present the student with a description of how to care for a child with a cleft palate. Actually, teachers can provide only guidelines for nursing care which are based on principles which they help students to learn. Instead of supplying a model and then observing how effectively the student fits her patient's care into it, the teacher must, after doing her best to make certain that the student possesses the necessary factual knowledge and technical skill, observe the process of her work with the patient as it unfolds. Since each situation is unique, she must observe what the student is doing at each stage of her care of the patient, evaluate with her its effectiveness, and help her modify techniques which do not seem effective. This approach to teaching fosters the development of students' individuality in giving patient care. It implies, however, that the teacher must restrain herself from developing the plan of care and then having the student implement it. It is natural that the nurse instructor should want to plan the care and then observe the effectiveness of her plan but, if she is to help her students develop individuality in their work styles, she needs to remember that while she can derive some satisfaction from guiding and participating clinically with her students, she can achieve it fully only by selecting certain patients with whom she herself can work, quite apart from her students' practice.

In order for the student to develop the self-confidence

necessary for growing independence it is important to handle situations in a way which assists her to be capable of the task to which she is assigned. If it is likely, for example, that a student will need help in getting a postoperative patient out of bed, it would be better for the teacher to be on hand to assist her when she is ready, instead of letting her proceed alone, find that she cannot manage, and then seek the teacher's help.

While the teacher cannot protect the student from failures and mistakes, and although the teacher recognizes that all of us learn from failure as well as from success, the difficulties and unpredictability of clinical work—coupled with human fallibility—provide amply for this learning. As far as possible, then, the thoughtful teacher will place the student in situations where she is likely to succeed, and where her skills are adequate to the demands made upon them. Otherwise the student tends to lose confidence and to "lose face" with patients and staff. These factors may cause her to start on a downward spiral of inadequate performance which elicits criticism and diminished esteem from others, and which, in turn, causes the student to be frustrated and angry with her own inadequacies and with the reactions of others to them; this reaction tends to lead to poorer, rather than improved performance. In addition, students who are assigned to tasks beyond their ability may begin to view the patient as demanding or unreasonable. The student referred to in the incident cited earlier might conclude that, had the patient been more cooperative, she could have succeeded in getting him up, rather than realizing that her own skills were not equal to the task.

Sooner or later the student does something better than her teacher. It may be a technical procedure at which the student has become more adept through more frequent practice, or a fresh insight about a nursing problem which makes the teacher wonder, "Why didn't I think of that?" Not to feel an occasional sting from these occurrences would be superhuman. But encouraging the growth of the student's individuality, even when this involves acknowledging that she has surpassed one's own skill on occasion, is part of helping her to develop her own professional identity instead of becoming a carbon copy of her teacher's professional attitudes and practices.

SUGGESTED READING

Cooper, Russell M., Ed. *The Two Ends of the Log.* Minneapolis, University of Minnesota Press, 1958.

Friedenberg, Edgar J. *Coming of Age in America.* New York, Random House, 1965.

Herge, Henry C. *The College Teacher.* New York, Center for Applied Research in Education, 1965.

Hill, Richard J. "The Right to Fail," *Nursing Outlook, 13*:38, April, 1965.

Hughes, Everett C. *Men and Their Work.* Glencoe, Ill., The Free Press, 1958, Ch. 7.

Mereness, Dorothy. "Freedom and Responsibility for Nursing Students," *American Journal of Nursing, 67*:69, January, 1967.

Smeltzer, C. H. *Psychological Evaluations in Nursing Education.* New York, Macmillan, 1965, Chapters 4 and 10.

4

Conducting Clinical Conferences

Clinical conferences are responding slowly and unevenly to changes in teachers' knowledge, skills, and convictions. In the past, the clinical conference frequently consisted of discussion of a patient's symptoms, diagnostic tests, treatment, and prognosis, with relatively little emphasis upon his nursing requirements. The leader might be the instructor, or a student; sometimes several students were asked to present material. One, for instance, might review the anatomy and physiology and the pathologic changes, another might present a summary of the patient's diagnostic tests, and contrast the results of these tests with normal values, and a third might discuss nursing care. More emphasis was usually placed upon medical aspects than upon nursing, and considerable time was spent in reviewing and reinforcing such factual knowledge as medical terminology.

Now, however, greater emphasis is being placed upon specific ways in which the nurse can help the patient. Students' experiences with patients are being used increasingly as the basis for the conference, thus helping to focus discussion upon particular patients and specific nursing actions necessary for their care. These experiences provide the raw material which the teacher uses to bring out the points she considers important. As is true in all teaching, the instructor's convictions are crucial in determining the way the data will be used. Her convictions are influenced by her knowledge and skill, however. The teacher who has a broad knowledge of pathophysiology, for instance, is likely to become convinced of its value and skillful in sharing this knowledge with students. If she believes that the chief value of clinical conferences lies in the opportunity for students to relate their knowledge of anatomy, physiology, and pathology to understanding the patient's disease, this will be the emphasis in her conferences. If she believes that medical information is truly the heart of the conference, discussion of differential diagnostic measures may take precedence over discussion of nursing care.

Sometimes, despite the teacher's earnest efforts to shift the focus of a conference toward nursing, or to broaden the scope of the conference to include more varied aspects of nursing, she discovers that the original emphasis persists, somewhat like a dandelion which refuses to be uprooted. It is easy, too, to fail to make the discovery that some topics are discussed repeatedly in various contexts, and others (which one had also planned to include) are not mentioned at all. A record of the topics discussed in conferences is helpful in locating these areas of over- and under-emphasis; students may be asked to take turns acting as recorder, or a tape recorder may be used.

USE OF MEDICAL INFORMATION AND OTHER RELEVANT THEORY

Sometimes, as a result of the effort to keep medical considerations from dominating the conference, medical information which is necessary to intelligent nursing care is omitted. For example, when considering the itchy rash of a patient with uremia, it is important to understand that this symptom is an accompaniment of the disease which is most effectively treated by ridding the body of retained waste products. This does not mean that students should disregard comfort measures like sponges and applications of calamine lotion, but having such knowledge does help them to avoid unrealistic expectations of the effectiveness of these procedures and to recognize the importance of such therapeutic measures as renal dialysis in relieving this symptom as well as many others from which the patient is suffering. Despite the concern to center the conference on nursing, it is essential that medical information which contributes to understanding the patient's nursing requirements be included. It is not a question of including nursing material instead of medical information, but rather of bringing various kinds of knowledge into a discussion which focuses on nursing.

For a conference discussion to move beyond a superficial level, the student must have knowledge of relevant theory to use in her consideration of nursing care problems. A systematic and careful study of the physiologic actions of various categories of drugs, for example, prepares her to consider the patient's response to medications and to understand the rationale of drug therapy, thus enabling her to help the patient learn about his treatment. Sometimes, however, enthusiasm for integration of concepts, and for emphasis on nursing, leads to lessened emphasis on factual knowledge. When a student is unaware of the purpose and procedure of a diagnostic test, she cannot explain it effectively to a patient, no matter how great her recognition of the teaching role of the nurse; nor can she help the patient learn how to follow a low sodium diet unless she knows what foods have high sodium content. The student who works with a patient who has paranoid delusions should carefully study theories concerning the causes of such reactions, which have been described by others who have worked with paranoid patients. Emphasis on theory imposes stringent requirements for disciplined study upon both students and faculty, but it yields dividends in more meaningful and thorough discussion of nursing and enables students to work more effectively with members of other professions.

THE USE OF DISCUSSION

Discussion of students' experiences with patients fosters an orientation of the entire conference toward nursing, while drawing on relevant information from biological and social sciences and from medicine. Variations in emphasis occur as a reflection of the particular interests of the students and faculty in the group, but the breadth and detail with which topics are pursued are determined largely by the purposes of the particular course. Students in a leadership course, for instance, are encouraged to concentrate on discussing ways in which they can care for larger groups of patients by learning to work effectively with auxiliary nursing personnel. This emphasis necessarily leads to less time being spent in discussing such equally important topics as instruction of diabetic patients, which is dealt with in another course.

The depth to which topics are explored is related also to the level of learning of the students; for instance, in the first clinical nursing course it may be sufficient merely to introduce the idea that families should be taught how to care for a patient when he returns home. The emphasis at that level is on the student's acquisition of beginning skills in providing certain aspects of care, rather than on teaching others. In a junior-year course, however, the student should be expected not only to recognize the importance of teaching the family, but also to formulate, carry out, and evaluate a plan for such instruction.

Students should be encouraged to share not only factual data during conferences, but also some of their feelings about the situations they encounter. Although talking about a nursing situation is especially meaningful for the student who cared for the particular patient under discussion, learning and emotional involvement are shared by the rest of the students. The immediacy and reality of the situation are conveyed by their recognition that the situation actually happened, and that it was experienced by a member of their group.

As students discuss their experiences, the material for learning is abundantly present, like a rich tapestry waiting to be unrolled. Some aspects of the material are more readily recognized by students than others. Part of the teacher's role consists in helping students to see some of the patterns and interrelationships of the tapestry's design which may otherwise elude them.

Recognition of the value of making patient-care situations the core of conference discussions sometimes leads to the temptation to dress up factual lectures in patient-centered garb. For example, if the purpose of a class is to learn about the therapeutic actions of Demerol, nothing is gained by using an introduction such as, "Mr. Brown fractured his hip. He is eighty-four years old, and lives with his wife in a suburban home. The physician ordered 50 mg of Demerol upon admission," and then proceeding with a discussion of the actions of Demerol without further reference to Mr. Brown.

THE ROLE OF THE TEACHER

The way the clinical conference is conducted differs with each teacher, since it reflects her particular way of working with students. One teacher may listen quietly and then guide and challenge her students with searching questions. Her thoughtful, supportive interest and her insightful questions seem to provide a subtle stimulus which enables students to plumb their own knowledge and insights, and to apply them to their care of patients. Another teacher may be all fire and sparkle—full of lively anecdotes from her own experience and eager to listen to students' accounts of their practice with patients. Her enthusiasm spreads to the students, who find each conference an adventure. Her approach stimulates students to think, too, and kindles their interest in the solution of nursing problems.

Challenging students to think

Whatever the teacher's particular style of leading conferences may be, her basic role is to challenge students to think—to make connections with past learning, and to push forward with consideration of aspects of a nursing situation which have escaped their attention or which have been superficially explored. After raising a question, the teacher should listen to what students say, and stimulate their thinking still further, rather than rejecting their ideas while trying to get them to provide the idea, or answer, which she has already formulated in her mind. She may tell one student after another, "No, that is not what I had in mind," until finally someone produces the desired reply. It is both stultifying and frustrating for students to realize that their ideas are not acceptable unless they correspond with those of the teacher.

Communicating constructive attitudes

As various topics are discussed during the conference, the teacher shares not only her factual knowledge, but also her attitudes and values, including her prejudices. It is important for her to be aware of her prejudices and to avoid passing them on to students. One teacher may, for instance, believe that surgeons are unmindful of their patients as people; another may think that elderly people are always suspicious. The teacher's comments can easily communicate these attitudes to students, so that they have the added burden of dealing with prejudices shared by the teacher, as well as with their own.

When a conference discussion touches upon a topic about which the teacher has a prejudice, it may be preferable for her to acknowledge her bias, thus allowing students to take account of it directly, rather than possibly having it conveyed indirectly or nonverbally. If, for instance, the question of patients taking their own medications is under discussion, the teacher who steadfastly believes that no hospitalized patient should be permitted to take his own medicine may tell the students how she feels about this practice, thus permitting them to consider her attitude while recognizing that not all nurses share her view. Since it is not possible, however, to protect students from exposure to prejudices, they should be encouraged to think about the effect which others' views may have on their attitudes. During conference a student may echo a staff member's comment, such as "The patient is very demanding." In such an instance it is wise to ask the student what evidence *she* saw of demanding behavior, thus encouraging her to substantiate her statements, and making her aware of the possibility that her attitude has been affected by that of the staff member.

Dissatisfaction with the way care is carried out on the ward often comes up during conference, and requires discussion. However, a nursing action which was performed with particular kindness and skill by a staff member or a student may not be mentioned, perhaps in the belief that since it is assumed that care will be given in this manner, no discussion is needed. Particularly in light of the widespread lack of satisfactory conditions for nursing care, it is important to help students recognize the instances in which care is handled capably, and which can thereby provide satisfaction and a pride in achievement. The teacher may have noticed, for instance, that one of the students held a five-year-old girl in her arms, rocking her gently in a chair as the child dozed until it was time for her to be taken to the operating room. Such compassionate and skillful care merits the teacher's comment, which should include both recognition of the student's contribution and explanation of the value of this care to the child. Students' criticism of the way others function may be frequent, and when valid, requires discussion, but it should not be allowed to overshadow the discussion of the actions which the student herself can take to help patients.

Providing a balanced perspective

The dramatic incident should not be allowed to overshadow the quieter but no less important challenges in nursing. Resuscitation after cardiac arrest is eagerly discussed by students, and can be capitalized on for the learning it provides. Equally important events, such as the gradual growth in ability of a child with cerebral palsy to speak intelligibly, or the gradual lessening of a schizophrenic patient's hallucinations, may receive little attention unless the teacher helps students to recognize and draw satisfaction from them. Beginning students, particularly, may believe that nursing consists of moving from one lifesaving drama to another. The serious accident with its many victims urgently needing emergency care compels attention in a way which a group of elderly depressed patients may not.

Some students believe that new learning comes only from caring for patients with different diagnoses. They may say, "Oh, I've already taken care of a patient with burns." Clinical conferences can help them discover the differences which make each nursing situation unique, and stimulate them to use each situation to increase their skill. Other students may develop the impression that their efforts are largely wasted unless there is considerable time for repeated contacts with a patient. Such an attitude denies the importance of quick, skillful, empathic care of people who are in trouble. Students need to understand that this kind of care is important in its own right. Clinical conferences can bring out the differences and similarities in nursing actions required in brief and extended care of patients, and can help the students to appreciate the value of both types of contact with patients.

Clinical conferences necessarily deal with many complicated situations—that of the unwed mother, the schizophrenic patient whose treatment has been long neglected, or the patient whose family have deserted him. There is sometimes a tendency to gloss over some of the unpleasant realities of such situations, perhaps to protect the students and oneself from considering them. The fact that the patient's family does not visit him at all may be briefly passed off, in conference, as "They seldom visit." This approach fosters unrealistic expectations among students, which can make their later practice in comparable situations very difficult. By passing lightly over some of the starkest dimensions of the patient's situation, the teacher may imply that all families are loving and supportive, that all mothers eagerly welcome their babies, and so on. Instead, students should be helped to recognize that the opposite is sometimes true, and be encouraged to discuss these painful and difficult realities straightforwardly, and to consider ways in which nursing intervention can help the patient.

Encouraging constructive criticism of nursing practice

In the course of the conference the ways in which care can be improved are pointed up, not in a humiliating or belittling manner, but as a challenge to the student to use her knowledge and skill more effectively. The teacher must convey her expectation that students are responsible and conscientious, and also her recognition that they are learners, and that the teaching-learning process requires that she make suggestions. All students participating in a conference should be encouraged to make suggestions as well, for the ability of students to provide constructive criticism of each other's work grows as they become more experienced. It is sometimes necessary, however, for the teacher to help students differentiate between stating opinions and making comments which are based upon application of theory relevant to a nursing situation. Mere stating of opinions can lead to arguments over whose opinion is correct and does nothing to help students develop understanding through application of theory to the practice of nursing.

The conference period is not a good time to discuss in detail the performance of a student who may be slower than her classmates in learning certain concepts. For instance, if a student, while going over interview data, shows that she repeatedly changes the subject which the patient has introduced, the teacher points out some of these instances during conference, but reserves some of them for individual discussion with the student. A repetition of "See, you did it again" can be humiliating for the student and tedious for the rest of the group. If, however, one of the other students calls attention to the student's changing the subject, the validity of her comment should be recognized so that she, and others in the group, do not become confused about the principles which are being taught. If a student makes a comment which indicates that she is having a seemingly inappropriate reaction to a situation, it is preferable to discuss the matter with her individually, after the conference. One student, for instance, may express a great deal of anger about the way a patient is behaving, but if neither the other students nor the teacher can see how it relates to the situation being discussed, then it is better to handle the problem with the student individually.

Sometimes the teacher decides that the time is not opportune to pursue a certain topic. Perhaps the discussion requires more time than is available, or possibly students need more background information before the subject can be handled satisfactorily. Whenever the teacher decides not to pursue a topic it is important to indicate this to students at the time, thereby helping them to recognize areas in which their knowledge is insufficient as a basis for nursing action. However, the teacher should state this in a way which emphasizes how the topic will be handled in the future, and not in a way which merely quickens students' insecurity about their knowledge and competence. For instance it is preferable to say, "Your conclusions about that are not correct, but it would be more useful for us to delay further discussion until you have studied the normal physiology," rather than to say, "Your conclusions are wrong."

Conference discussion can be useful when an error has occurred, such as the administration of an incorrect medication. News of such an incident usually passes quickly from one student to another even before the conference commences. These furtive discussions often evoke insecurity among the students, who realize that they, too, could make a similar mistake, and who wonder how the situation will be handled. The question for the teacher to solve is not *whether* the other students should be involved, but *how* they should be involved. Before the conference begins the teacher should talk with the student who made the error about whether the incident will come up, and if so, how the discussion will be handled. It is advisable to ask the student ahead of time whether she is willing to share this experience with her classmates, so that they, too, can learn from it. If she feels that she cannot discuss the incident before her classmates, her decision should be respected. But if she is willing to share this experience, she leads the discussion about it. Thus, the student is in control of the way the topic is handled, and she knows that the teacher will not "spring" some possibly humiliating question about the error in front of the group.

Talking about the mistake during conference can benefit the student who made the error, and her classmates as well. The incident is discussed openly, thus avoiding some of the misinterpretation which is likely to occur when the account of an error travels to students via the grapevine. By admitting her fallibility and by using the experience in a way which helps her classmates to learn, the student who made the error earns the respect of her classmates. The teacher is given a chance to show her support of the student and her belief that errors can be used to foster learning. Observing the teacher's response to the incident can help other students to feel more secure about reporting and discussing any mistakes which may occur. They have an opportunity to voice some of their worries about making mistakes, and to consider ways in which they can minimize chances for error. Discussions can help students to realize that responsible, capable people can make mistakes, and that, if they do, they should not be automatically considered unworthy of respect. The importance of safeguarding the patient by prompt reporting of errors, and correction (whenever possible) of the circumstances which led to the mistake, should be stressed in relation to the incident. The teacher's best procedure in such a case is to seek a useful middle ground between the extremes of a punitive attitude and one which is so casual that the seriousness of errors is underestimated. Encouragement of openness in reporting and discussing errors does not imply laxness or neglect of the patient's safety. On the contrary, discussion lessens the dangerous problem of concealment of mistakes due to fear of punishment, and it can help students develop a conviction that the occurrence of an error—their own or someone else's—provides an opportunity for learning.

Maintaining interest

The tone of the conference fosters or impedes discussion. Sometimes the reasons for a certain tone or atmosphere which dampens students' participation are stubbornly elusive. Everything may have ben planned to encourage exchange of ideas. The room is quiet and comfortable, chairs are arranged in a circle; there is plenty of time. Nevertheless, the ideas do not seem to flow. No one seems to want to say anything. Tedious silences are broken by routine reports each time the teacher calls on a student. In another conference, the same group of students participates eagerly, asking questions of the teacher and of each other, debating nursing problems with a zest which makes it seem impossible that they are the same group who attended the previous conference. Every teacher has had such experiences. Sometimes a conference just does not seem to "click." Although it may be hard to pinpoint the reasons, it is important to try to do so. Was it just plain weariness which spread through the group shortly before the end of the term? Had most of the students quietly decided not to participate because one student had monopolized the last three conferences? Was it because the teacher, at this particular conference, was conveying an attitude of "Why don't you know that?" In addition to her own careful reflection about the conference, the teacher may find listening to a tape recording of it helpful in her search for the difficulty. If a colleague was present, she can be asked to help in finding the cause. Students also can be asked what they think the problem is. Studying the tone of conferences, and considering factors which may impede discussion, can help one to reduce the number of these mutually unsatisfying experiences for the students and teacher. However, because it is not always possible to understand all the complex influences on the conference situation, the teacher, no matter how expert, must be able to accept the occurrence of an occasional conference which seems to fall flat for reasons which at the time defy understanding.

Dealing with emotionally charged material

The teacher of nursing is more likely to deal with emotionally charged material than is, for example, a teacher of typing, mathematics, or physics. Because of the universality of human problems, students and their teachers may have experienced situations similar to some of those which come up during clinical conferences. This presents not only the possibility of a richer and more meaningful discussion, but also of difficulties in dealing with the subject. Some of these topics resemble lightning which, while it briefly illuminates, can also strike too close, and startle by its unexpectedness. Such an instance occurred during one conference when a student suddenly burst into tears and left the room. Talking with the student later, the teacher learned that the patient's circumstances were so like those surrounding the illness of the student's husband that she was overwhelmed. The reaction of one member of the group may be triggered by an apparently casual remark, or one which is merely a statement of fact. For instance, to one student's query, "Can people die from an asthmatic attack?" the instructor replied "Yes." This interchange, though brief and factual, was distressing to another student whose father had recently died in an asthmatic attack, and who knew only too well the answer to her classmate's question.

When the teacher notices that a student is becoming upset, it is important to help her retain her composure, if she can, without implying that a show of emotion is shameful or unacceptable. Sometimes introducing a concrete and very factual topic, such as the action of certain drugs a patient has been receiving, can help the student maintain her poise and continue to participate in the conference. Later the teacher can talk individually with the student and, by showing her concern, give the student an opportunity to discuss the difficulty if she wishes to.

It may also happen that a situation being described during conference so closely resembles the teacher's own experience that she is momentarily at a loss for words. She may quickly recognize that the topic is touching a personal concern or experience, and because of this, be able to enrich the discussion by sharing not only factual knowledge, but a depth of understanding acquired by living through a similar experience. These are moments when a conference becomes suddenly more alive; not only is the teacher giving more, but students are listening with greater awareness and responsiveness.

Two less effective ways in which the teacher may respond to such situations include lapsing into an emotional catharsis concerning her own experience or, at the other extreme, handling the discussion in a coolly detached manner. The former is inappropriate to the situation, and interferes with the instruction which should be taking place. The latter enables the teacher to present the necessary material, and she may do so quite effectively, but if she uses this self-protective device frequently, she may communicate to students an aloofness and remoteness from clinical problems. This can happen, for example, when a teacher who is deeply distressed by the helplessness of paraplegic patients jokes too much about their problems, and does so in a way which conveys coolness and possibly even a lack of respect for this group of patients. Another teacher may discuss the foibles of the hospital "system" with such a detached and, at times, satiric manner that she conveys a lack of concern for patients, despite the veracity and occasional brilliance of her remarks. The way the teacher copes with emotionally loaded material is not wholly a matter of deliberate choice. Sometimes a topic is too painful for her to handle except in a highly intellectualized way. However, the more aware she is of the way her response to certain topics either limits or enhances her ability to deal with them during clinical conferences, the more likely she is to be able to use her own emotional reactions to enrich, rather than to detract from, her teaching.

Occasionally a student's emotional reactions to conference topics are linked with concern over her own health, or that of a family member. She may silently absorb information about the illness, or she may closely question the instructor about symptoms, treatment, and prognosis. She may ignore the differences between the patient's situation and her own, and concentrate on the similarities until for example, she visualizes herself as being promptly and totally incapacitated by rheumatoid arthritis, although she has had only one very mild attack of this disease. The teacher, of course, has no way of knowing what inferences concerning their own health students are drawing from conferences, unless students bring them up. A student may mention such a concern to the teacher after the clinical conference, or during a scheduled individual conference, and this gives the teacher an opportunity to point out that the student's health problem, despite its apparent similarities, is not necessarily the same as that of the patient, and to suggest that the student talk with her physician.

It is not unusual for a student to recount in some detail, during a clinical conference, an experience with illness which she or a member of her family has had, and to attempt to engage the teacher in a discussion of why certain treatments were used rather than others, and what the outcome can be expected to be. Occasionally the teacher finds herself drawn into this type of discussion before she realizes what is happening. What at first appeared to be a useful illustration of a topic from the student's own experience becomes an attempt on the student's part to verify whether treatment was adequate, or whether a prognosis is favorable. Such discussions can usually be terminated by gently reminding the student that this kind of query should be addressed to the physician, who has the relevant information concerning the situation and is, therefore, the appropriate person to give the guidance which she is seeking.

Material presented at clinical conference can also have an emotional impact upon a student by its particular relevance to a problem which she is facing. Such a fortuitous meshing of a topic with a particular student's concerns can occur in any course, or in several courses. One student, for instance, was concerned about the problem of honestly being oneself, without pretense or apology. In an art class she had heard the teacher say, "Materials should be respected for what they are, and not made to resemble something they are not. Formica is a beautiful material, but it is different from wood. To try to make formica look like wood destroys its beauty by trying to make it appear to be something which it cannot be." The student drew a parallel between this comment concerning respect for the particular characteristics of materials and one she had heard in a clinical conference a few days previously. The teacher had suggested that the fact that one family member openly expressed grief at the terminal illness of a patient, while another maintained composure did not mean that either reaction was good or bad, or that the nurse should attempt to elicit an expression of grief from the composed person, or scold the more demonstrative one but that, instead, she should respond to each of them differently and according to their reactions.

Such relevance of a discussion to a student's personal concerns can be the basis for the unexpected and seemingly inexplicable enthusiasm or gratitude which a student expresses about a certain conference. "That really came home to me," or a similar comment, can be the way a student conveys her feeling about a discussion that touched upon some personal concern, releasing a fresh insight.

TYPES OF CLINICAL CONFERENCES

Pre- and post-experience conferences

As their names imply, these types of conferences are intrinsic parts of the process of clinical teaching, and therefore they will be discussed at the appropriate times in Chapter Five. We should, however, give attention here to a third and very useful type—the interdisciplinary clinical conference

Interdisciplinary conferences

This kind of conference can assist students to develop a clearer understanding of the contributions of others to patient care, and of the role of the professional nurse in relation to other health workers. It also gives the student practice in planning with others for the improvement of patient care. Particularly helpful is the opportunity for students to present nursing observations and plans for discussion, to clarify and defend these during the conference, and to recognize the interrelationship between these plans and the plans of other professional persons for the care of the patient. How to help a paralyzed patient begin to use crutches, for example, can be discussed by the physical therapist, the nurse, the physician, and the clinical psychologist. As a result of such discussion the students can gain heightened appreciation of the importance of coordination in plans of care, of communication among all who are working with the patient, and of the areas in which the discussants' various roles overlap. This type of conference experience may be scheduled as a part of the students' educational program, or it may be a regular part of the agency's plan to which students are invited. Regardless of how the conference is initiated, it serves both the patients' needs for coordination and planning of care, and the educational needs of all who participate. Attendance at interdisciplinary conferences is an excellent way for students to learn about the mutuality of interests between professional groups.

When the teacher initiates this type of conference, she should carefully orient each participant beforehand to the background of the students who will be present, and plan with them the way in which the conference will be handled. Such preparation ahead of time can help those unaccustomed to such conferences to understand their purposes, and not to interpret them as occasions for presenting lectures on disease. Although this type of conference experience is essential for all students, it is particularly so for those in collegiate programs, whose contacts with other professional groups may be limited due to the shorter time that they are present at the agency.

The role of the teacher in such a conference may be that of moderator (in which case a student assumes the responsibility for discussing the patient's nursing care), that of the nurse-participant (thus providing a role model for students), or a combination of both. Her decision about how she will function in the conference is affected by the level of the students' learning and by their previous experience with interdisciplinary conferences. In planning for the first such conference with beginning students, it would be appropriate for the teacher to assume major responsibility for the discussion of the patient's nursing requirements, with the participation of the students who have cared for the patient. However, senior students who have had considerable experience with interdisciplinary conferences can be expected to present the major discussion concerning nursing care, and then the teacher becomes a participant and moderator. The student who assumes major responsibility for discussing the patient's nursing requirements should have the teacher's guidance in preparing for the conference.

Important prerequisites for fruitful interdisciplinary conferences are a shared body of factual knowledge, as well as clarity among various members concerning their own professional role with patients, and appreciation of the ways the roles of others differ in their orientation and emphasis. Sometimes so much stress has been placed on the process of conducting clinical conferences (such as whom to invite, and how to foster discussion) that insufficient attention has been given to these essential prerequisites. Unless the participants share a certain body of content, such as knowledge of the patient's condition and treatment, and understanding of the disease process affecting the patient, the conference may readily turn into a lecture during which one member imparts information to the others, or the necessary group planning may be impeded because those present do not share sufficiently the fundamental knowledge which serves as a basis for useful discussion and planning of the patient's care.

Clinical conferences provide opportunity for the intelligent application of theory to practice. Effective leadership of these conferences stems from the teacher's firsthand knowledge of the nursing situations being discussed, her ability to help students utilize relevant theory in their study of nursing problems, and her sharing of interest and concern for patients.

SUGGESTED READING

Lister, Doris W. "The Clinical Conference." *Nursing Forum*, 5:84, No. 3, 1966.
Newman, Margaret A. "Identifying and Meeting Patients' Needs in Short-Span Nurse-Patient Relationships," *Nursing Forum*, 5:76, No. 1, 1966.
Palmer, Mary Ellen. "Nursing Care Study Brought up to Date," *Nursing Outlook*, 12:36, June, 1964.
Shetland, Margaret L. "Teaching and Learning in Nursing," *American Journal of Nursing*, 65:112, September, 1965.
Zelko, M. P. *Successful Conference and Discussion Technique.* New York, McGraw-Hill, 1964.

Part II

THE PROCESS OF
CLINICAL TEACHING

5

The Process of Clinical Teaching

The reality of the clinical situation is its strength. It is easy to forget this when confronted by the imperfections which are also part of most clinical environments—the equipment which is lacking or broken, the crowding of patients into clinic waiting rooms, and so on. But no classroom discussion can convey to the student the joys and satisfactions of nursing as vividly and meaningfully as holding and feeding a sick baby, seeing the pride in the face of a new father, or easing a postoperative patient's pain. One student worked day after day with an aphasic patient helping him learn to talk again. The day he spoke his first word, she said, with tears of happiness in her eyes, "He said his dog's name, and I feel as though I've received a lovely present."

In the clinical situation the student confronts some of the challenges and difficulties of dealing with real patients, and experiences the painful recognition that not all of her patients will recover and that not all of them can follow instructions concerning their health. Nevertheless, the harassing and irritating experiences provide excellent opportunities for learning. It is no service to a student to be led to believe that all physicians and nurses are invariably kind and explain matters thoroughly to their patients, or that all the clean linen one needs is unfailingly available. By studying her own reactions and learning how to cope with situations involving apparent lack of consideration for patients, or lack of necessary supplies, the student will be better prepared to handle them effectively when she has graduated.

After a lengthy discussion of educational issues with her colleagues, many a teacher has said to herself, "Somehow the theories can be sifted and understood better, when one returns to the solid reality of a sick patient and a student who is learning to care for him." Without frequent exposure to the clinical situation it is easy for teachers to magnify certain problems beyond their actual importance, and to miss others entirely.

The opportunity which the student has to relate classroom learning and clinical practice is of inestimable value. What teacher has not turned to a student whose recent clinical experience illustrates a point under discussion, and invited her to share this experience with the group? The incident is made more meaningful for all the students because it is something that has actually happened to one of their classmates. The most effective learning can occur when student and teacher together confront the reality of the patient's situation, and draw from it the implications for nursing action.

Freedom to use clinical resources to further students' learning has been increasing in nursing programs, and particularly in those conducted by colleges and universities. This freedom demands that the teacher face the difficult task of deciding what kinds of clinical learning experiences the student requires in order to accomplish certain objectives. To say this is easy; to do it is not, because it requires evaluating many of the usual practices in schools of nursing. Some experiences may be eliminated only to be reinstated later when one realizes that they are important after all; some valuable opportunities may be ignored because they conflict with established practices one is not ready to question.

Suppose, for example, that a student is studying about the care of patients who require breast surgery. Her patient has gone to the operating room for removal of a tumor which may prove malignant. While the patient is in the operating room the student makes the patient's bed, and then helps her classmate get a heavy patient up. Would it not be more useful for her to be with her patient in the operating room, to experience some of the suspense which occurs as the surgeon awaits the pathologist's report, and to observe the extent of the surgery which is actually performed? If the student has prepared other beds for postoperative patients and has helped other heavy people out of bed, foregoing these experiences in order to go to the operating room seems preferable. Then why did she not go? Perhaps because it is the teacher's duty to see that the assignment is completed before the students leave for class and, today, no other student is free to make that particular bed. The teacher does not have time to do it, nor can the staff take it on at the last minute. (It is easy to consider these problems in retrospect. Upon reviewing this situation it is apparent that, had it been clarified to the staff that the student's assignment was to give immediate preoperative care and go to the operating room with the patient, they could have planned on making the patient's bed.) Perhaps it never occurred to the teacher that the student *could* go. The teacher may be granted considerable freedom to select clinical learning experiences, but it often takes time before she feels free to use this freedom effectively.

Suppose a student has been assigned to care for an arthritic patient for three days. The patient is to be discharged with a referral to the public health nurse. Would it be useful for the student to accompany the public health nurse in order to see the patient in her home environment, and to observe how her nursing care there will be similar to and how it will differ from the way the nursing care is approached in the hospital? Yes, this sounds like a fine idea, but what about the student missing the morning's experience at the hospital? She cannot make too many home visits and still master the many skills needed to care for a variety of patients. Actually, if the students' clinical experience is carefully planned within the total program, as well as within each course, the possibilities for varied experiences are much greater than one might suppose. Can the student be assigned to work with a child in traction, if not with an adult, or vice versa? Can two students be assigned to care for a heavy adult in traction, thus sharing the available experience as well as solving the problem of finding help when two people are required for some aspects of care?

PLANNING CLINICAL ASSIGNMENTS

The student's future competence as a practitioner of nursing depends, to a large extent, upon the quality of instruction provided during clinical practice periods. It is then that students can apply and refine concepts presented in class, and develop the skills and judgment which will be required of them as practicing nurses. Responsibility for planning the clinical assignment rests squarely with the teacher. Just as control and financing of the educational program in nursing affects the way in which the entire program will be carried out, control of the assignment is a crucial factor in determining the manner in which clinical laboratories will be conducted.

Control, in this sense, emphatically does not mean that personnel in the agency are not acquainted with the purposes of the clinical practice periods, or that they do not make suggestions concerning the kinds of learning experiences a particular unit may contribute to the fulfillment of these purposes. It does mean, however, that the decision concerning which aspects of patient care, which patients, and how many patients each student will be assigned rests with the teacher.

To say that the teacher controls the clinical assignment does not imply that she is impervious to the influence of the expectations of others such as the nursing staff and physicians. However, as their expectations become more similar to hers (for example, after a program has become established in a clinical setting), the student is exposed to fewer conflicting views concerning what she should be doing, and the teacher is freer to develop plans for clinical learning experiences. Some of the energy she previously used to interpret the purposes of the program and to lessen misunderstandings can be channeled into her teaching.

Stating objectives clearly

Educators have consistently placed emphasis upon the necessity for clarity of objectives and the selection of experiences which are appropriate in achieving these objectives. This sounds not only logical, but easy. Why then is it so hard to do? One reason is that it is difficult to state an objective clearly. Vague generalities concerning "care of the whole patient" and "meeting patient needs" abound. In contrast, the statement that an objective of a clinical laboratory is to learn to bathe another person seems, perhaps, commonplace. Nevertheless, it has clarity, and gives direction to selection of experiences. Stating the objective is essentially an intellectual process. Once the teacher arrives in the agency, she is assailed by the tradition of her own clinical experience as a student (which may or may not differ from the experiences she plans for her students), and by the expectations of staff, patients, and students.

The student is assigned to bathe a patient, but the staff explain that whoever gives a morning bath must also care for the patient's flowers, deliver his mail, and take him to any scheduled appointments. The physician believes this is the time for the student to learn to distinguish various heart sounds, and proceeds with that lesson. The patient states he has been bathing himself, although he knows he is not supposed to. Now where is that carefully planned learning experience, based upon that well-formulated objective? Provided the teacher has thought these matters through, her success in implementing her ideas will depend upon her conviction concerning their value, and her ability to deal tactfully but firmly with others who do not share her views. Lack of conviction on the part of the teacher can lead to the student's being pulled in so many directions at once that she is unable to concentrate on learning to give the bath.

The teacher must be guided not only by the student's need for education, but also by the patient's requirement for continuity of nursing care, and his need for a sense of security in the quality of care he will receive from the student.

Matching the right student with the right patient

The teacher needs to take account of the different abilities of students when planning their clinical assignments. The patient who challenges one student may overwhelm another. In addition, some of the subtle factors which affect every clinical teacher's planning of assignments must be recognized. There is a difference between challenging the able student and using her to bail one out of an unwise assignment. Perhaps the patient's care plan presents problems which no student at this level should be expected to handle. Conversely, the less able student should be helped to move toward the level of performance expected of all students in the group, and when such help is not effective, she should not continue in the course; consistently lessening the demands of her clinical assignment is necessary to protect patients, the student herself, and the teacher, but it means also that the same criteria for evaluation of clinical performance are not being applied to all students in the group. No magic formula exists for coping with these subtle influences upon assignment planning, except that the teacher can, perhaps, sharpen her awareness of these influences when making clinical assignments, and clarify more fully the level of performance expected of students in each course and at different periods during the same course.

One important facet of planning assignments involves recognizing what else must be dealt with in the situation, besides the particular experience being planned—giving a bed bath, for example. The student may be assigned to give care to a patient on bed rest following myocardial infarction. He needs to be bathed, he is in no acute physical distress at the time and, if special equipment is not needed, he may seem like an ideal patient for the beginner. However, one must recognize that, by the nature of his illness, his condition is likely to change suddenly, and the nurse who is with him needs to be constantly alert to early signs of such change and prepared to act quickly should they occur. The ninety-six-year-old patient who moves slowly and with difficulty, and whose verbal responses are slow and wandering, also may require more nursing skill and confidence than the beginner possesses. The ideal patient for a student's first experience in giving a bed bath is hard to find, particularly in the general hospital, because most patients who would be suitable for the beginning student are permitted to assist with their own bed baths, or to use the tub or shower.

Although the word "laboratory" is often used to designate clinical experiences for students, the nursing laboratory is by no means analogous to one held in courses in the physical sciences. The most experienced and skillful teachers cannot foresee just how the assignment of a particular student to particular patients will work out. Who can predict all the changes which may occur in a patient's condition, or that a patient's absence from the ward for tests or treatments may be twice as long as expected? The responsibility the instructor assumes is demanding, emotionally and physically, for the combination of a sick patient and an inexperienced student often taxes one's skill, both as a nurse and as a teacher. It also places a premium on patience, alertness to the needs of both patient and student in the nursing situation, and judgment about how much latitude in decision-making to allow the student, both for the patient's welfare and that of the student.

The amount of time spent in clinical laboratory has diminished markedly in undergraduate nursing programs, thus making it especially necessary to use the available time as effectively as possible. The teacher must carefully consider, beforehand, the purpose of the particular laboratory period and the kinds of experiences most likely to fulfill this purpose. Perhaps the purpose is to provide practice in preparing patients for surgery, observing surgery, and giving postoperative care. A group of preoperative patients is selected, with the expectation that each student will care for a particular patient on three consecutive days: the day before the patient's surgery, the day of surgery, and the day after surgery. Next comes the task of deciding which student should work with which patient. Certain patients in the group will provide more difficult challenges than others; some will require more assistance with personal hygiene, or with such postoperative nursing measures as encouraging deep breathing. Which student in the group has had less opportunity than her classmates to give physical care? Perhaps she should be assigned to care for the elderly lady whose hip is to be pinned, and who will be in bed on the first postoperative day. The student, as well as the teacher, should be aware of the purpose of the assignment. She should know, for example, that one expectation is that she will observe the breathing of the patient in pulmonary edema or, if the patient happens to be diabetic, that she will be expected to measure and inject his insulin.

What kind of information does the teacher need about the patient before planning the assignment? Certainly she needs to know the diagnosis, but this is only the beginning. The patient about to have a dilatation and curettage may be suitable for the inexperienced student. But suppose the patient is a young, apprehensive woman who is suspected of having cancer, and who knows that, depending on what is found during the operation, she may have to undergo a hysterectomy. Because of her requirement for skillful emotional support, this patient may not be a good choice for the inexperienced student. Who are the other patients in the room? However convalescent the assigned patient may be, it would not be wise to assign him to an inexperienced student if the patient's roommate is dying and grieving relatives are at his bedside. It is easy to underestimate the impact on the student of other events on the ward. Having planned the assignment, one may think that the student's experience will be limited to the particular patient selected, rather than being affected, as it surely will be, by a vast variety of circumstances, such as overhearing conversations among staff at the nurses' station, or observing the behavior of other patients and their families.

How can the teacher know all the factors which may affect an assignment? She can't, but the more of them she can recognize, the better. If time and distance permit, it may be desirable to visit the unit and select the patients the day before the scheduled laboratory. If this is not possible, a telephone consultation with the head nurse may suffice, particularly if she is accustomed to working with the students and teacher. Least desirable is leaving a note for the head nurse, saying, "Six cardiacs, please" or "Six convalescent patients, please."

Planning for continuity of care

Although providing for continuity of care is important, it is often sacrificed in order to give the student variety in patient contacts. During her first experience with a patient the student may observe and deal with certain aspects of his care, and become aware, with the help of the teacher and members of the agency staff, of other aspects. On the second day she can provide more complete care, and become aware of additional factors requiring consideration; she is then familiar, for instance, with the bound of patient's pulse, the redness of the spot on his back, and how tight to make the elastic bandage. By the third day she may find a way to talk with the patient that helps him understand his illness. Opportunities for continuity of care are limited in many general hospitals due to the turnover of patients and the limited hours students are present, and to the need for students to have contact with different types of nursing situations. At some point in the curriculum, it seems desirable to include opportunities for care of long-term patients, so that the student can work with a particular patient and his family over a period of time. If one mark of the professional nurse is her ability to assume responsibility for a patient's care over time, and to deal effectively with the many facets of the nursing situation as it unfolds, the student must have opportunity for this kind of experience. Otherwise, this aspect of nursing is dealt with only through verbal generalities about the value of long-term planning for the patient's care.

It is helpful, too, for the student to see some of her patients in more than one setting, so that she can develop an understanding of the importance of continuity of care. Caring for the patient in the hospital and then visiting him at home, or caring for a patient in an acute hospital setting, and later in a convalescent-care setting, can help the student understand some of the real problems which arise in relation to continuity of care and some of the ways in which the nurse can help with these problems. Such experiences also help her to recognize the manner in which different settings affect the way care is given.

Providing adequate supervision

The clinical assignment must not be more extensive than the teacher can supervise within the allotted time. Nothing could sound simpler or more obvious, but the application of this concept is often extremely difficult. The natural desire not to pass up a learning experience which may be hard to find later can make the temptation to over-assignment almost irresistible. One may also assume that each experience will move along at an optimum rate—all the equipment readily at hand, all the students eager, alert, and confident, and all the patients fitting into the plan like pieces in a puzzle. Such is usually not the case. Many a teacher can blushingly recall a day when, for example, three students were late to class, a fourth did not complete the patient care assigned, and the teacher arrived at faculty meeting when it was half over. When planning assignments, one must make allowances for the slowness of neophytes and the imperfection of supplies, and for such unscheduled necessities as the patient's need to stay on the bedpan longer than usual.

No one can predict exactly how long a certain assignment will take a particular student, and it is essential to have some additional experiences in mind which can be used for students who have time for them. These should not be of the "busywork" variety, which merely occupy the student and require little supervision—running errands or filling water pitchers are in this category. On the other hand, the idea (held by some teachers) that students should never help members of the staff seems unfortunate, since such activities can benefit the student by adding to her learning and can also help to improve staff-student relationships. The teacher can discuss with the head nurse or team leader the areas in which a student's help would be useful and then select such an activity as feeding a helpless patient or talking and playing with a frightened child, which would be meaningful to the student as well as helpful to the staff.

If one student requires a great deal of supervision in carrying out a particular assignment, other students must be assigned so that they require less of the teacher's time. Perhaps the challenge lies not so much in acknowledging the principle as in the development of the ability to make educated guesses concerning the problems inherent in particular nursing situations and the resources of particular students for dealing with these problems. The difficulty stems not from deciding ahead of time that one can be three places at once, but from inaccurate predictions as to how much direct supervision various students in the group will require.

The difficulties inherent in an assignment are hard to evaluate in advance. Most readily apparent are the technical aspects with which the student is unfamiliar, but even with these it is hard to predict how much difficulty a student will encounter. She may easily master such comfort-giving techniques as the back rub but find it hard to learn to perform such pain-producing procedures as the administration of injections. One student may adapt quickly and easily to care of children, but find it difficult to care for the aged. The patient's diagnosis may make it difficult for a student to work with a particular patient. Perhaps this is because the very word cancer frightens her, or because paralysis with its helplessness touches on a deep fear of hers.

Coordinating clinical and classroom learning

Coordination of clinical and classroom learning is facilitated by formulating units of study which are sufficiently broad that one is reasonably certain to find suitable patients for students' assignments. For example, a study of the care of patients who have disturbances in locomotion can encompass fractures, amputations, multiple sclerosis, and many other conditions. In contrast, it may be difficult to find patients to assign during a study unit dealing solely with patients who have fractures.

Following the students' clinical experience in caring for particular types of patients, such as the patient in traction, or the amputee, subsequent class discussions can be illustrated by the students' experiences. Such a plan requires that the students know which patients they will care for at least one day in advance, to permit them to look up requisite information. It also necessitates sufficient numbers of faculty to demonstrate unfamiliar equipment and discuss with the student the material she has studied prior to her clinical practice.

Fostering smooth relationships with the staff

A thoughtful approach to the making of clinical assignments can foster smooth working relationships with the nursing staff. An assignment which is planned a day in advance and agreed upon by the head nurse and teacher, and which is clearly posted, shows respect and consideration for the fact that staff must plan assignments for the nursing personnel on the ward. Tailoring the assignment to fit, as well as possible, into the laboratory period helps avoid the problems created for staff when students leave the ward before care of their patients has been completed.

*Recording results to help with the
planning of future clinical assignments*

It is essential that the teacher keep records of the students' clinical assignments. Sometimes saving the assignment sheets and adding notes to them is helpful in jogging one's memory concerning what aspects of nursing care each student has dealt with previously. Whatever the mechanics of their maintenance, the records should go beyond the checklists of procedures sometimes used by students and teachers. The items recorded should reflect the purposes of the course. If one goal is to help students learn to work with toddlers, some record of how often each student has been assigned to a toddler will be necessary. How detailed records should be depends to some extent on the size of the group of students, the number of specific items one is observing, and the adequacy of one's memory. Although a narrow concentration on procedures is not desirable, they should not be overlooked when reviewing the student's past experience and planning her future clinical assignments. A lofty disdain for procedures, coupled with the objective of preparing *practitioners* of nursing is incongruous. In addition to guiding the individual teacher in planning, records can facilitate coordination of planning among various faculty members. If a certain student has never cared for an adult who has a cast, for example, maybe she can have this experience during her course in nursing of children.

A lingering concern with having the student master many procedures at the beginning of her nursing program persists. In the past, this concern had a very realistic foundation, because in some schools the student had the help of a teacher for only the first six months, or perhaps the first year, of her program. After that she was often expected to take charge of a ward at night or in the evening, and there would not be time then to show her how to give an enema or do a catheterization. Now that many schools provide the student with the assistance of clinical teachers throughout her program, it seems more appropriate to space the learning of nursing procedures more evenly. Difficulties in scheduling certain types of experience also make this necessary; for example, catheterization is a procedure which has become rare in many services.

When, and to what extent, should the student select her own assignment? During the early part of her program she does not have enough familiarity with nursing, and with what there is to be learned, to enable her to make her own selection. During this period, however, the student should be aware of the purposes of each assignment, alert to the kinds of experiences she has and has not had, and sensitive to the extent of her self confidence when in various types of nursing situations. Sometimes a student performs care so competently that the teacher thinks she should have a different assignment the next day. The student, however, may state that although the patient's care progressed satisfactorily, she herself was tense and lacking in confidence. Thus, even the beginning student can participate in the planning of her assignments. The senior student, by contrast, should assume more responsibility for identifying topics of particular interest to her, as well as areas with which she needs additional experience, although she, too, needs the teacher's guidance as she selects certain aspects of nursing care for study. The process of selecting a patient or nursing problem for study is in itself important, as it helps the student to identify her interests and sharpen her perception of the possibilities for learning inherent in a given situation.

It is no exaggeration to state that the skill with which the assignment is planned determines, in large measure, the value of each clinical experience. Effective planning of clinical assignments can challenge the student without overwhelming her, enable her to synthesize and utilize concepts and skills more effectively than she has previously, and stimulate her to seek creative solutions to nursing problems. Thus it can be seen that planning clinical assignments is the crux of preparation for clinical instruction. To do so with clarity concerning the purposes of clinical practice periods, independence and conviction about which experiences are appropriate to carry out these purposes, and flexibility necessary for functioning in a clinical agency establishes the framework for effective clinical teaching.

THE PRE-EXPERIENCE CLINICAL CONFERENCE

The principles upon which a successful clinical conference is based (see Chapter 4) are immediately relevant to the pre-experience conference which prepares the student for clinical practice or observation, whether it occurs in a hospital, a public health agency or other community health facility. This conference provides opportunity for students and faculty to take stock of the situation with which they are about to deal, and to determine how they will proceed. Students have a chance to ask questions, some of which can be answered during the conference, others during individual discussion after the conference. The teacher is enabled to gain a clearer idea of how effectively students have already prepared for their assignments, and to note areas in which they will require assistance or further study before they can proceed to give care to the patient. Needs of various students for assistance during the laboratory period can be predicted, and plans made so that such assistance is not sought at the last minute. The student who must get a heavy patient out of bed, for instance, may be advised to contact the orderly promptly, and arrange a mutually convenient time when he can assist with this care. The teacher can plan with students for the periods when she will be with them to help with certain aspects of care. Such foresight can help avoid the situation in which all students in the group are seeking the teacher's assistance at once, because their requirements for supervision and instruction were not anticipated.

The pre-experience conference does not replace the orientation given by the head nurse or the team leader. Orientation is needed by faculty and students to ensure continuity of patient care. Since the head nurse or team leader usually has the most up-to-date information concerning the patients, and since the patients' care will be continued by the staff when the student leaves the ward, it is essential that the orientation be provided by a member of the nursing staff. This also leads to more meaningful contact between the students and the head nurse or team leader. The head nurse, however, does not ordinarily know that a particular student has never changed a colostomy dressing before, or that her classmate has never measured and injected insulin. These are the teacher's concerns. The pre-experience conference led by the teacher helps students to understand the material presented by the head nurse or team leader, and to recognize how they can implement her recommendations, such as "Encourage fluids" or "Help him understand his diet."

All students should attend both the pre-experience conference conducted by the teacher and the orientation to patients given by the head nurse. Sometimes students stay only as long as their own assignment is being discussed, thus limiting their familiarity with the assignments of other students. By listening to the head nurse's orientation to each patient, and to all of the discussion during the pre-experience conference conducted by the teacher, students have an opportunity to become aware of the kinds of nursing situations their classmates are dealing with, thus broadening their own knowledge and alerting them to the topics which will be discussed in the post-experience conference. Students' ability to attend to aspects of the discussion which do not apply to them varies. A student may be very tense about giving an enema for the first time and not hear anything during the pre-experience conference which does not have to do directly with her own assignment. However, for most students, most of the time, giving attention to what their classmates are doing is a useful device for broadening their learning. A brief visit by the group to each of the assigned patients immediately after the morning orientation is helpful in making the post-experience conference more meaningful to all students. Such visits allow students to make observations about the patients who will be discussed later; certain pertinent observations can be pointed out at this time, such as the appearance of a decubitus ulcer, or the limited motion possible to a patient whose hands are crippled from arthritis. (These visits are carried out with careful consideration of what is said in the patient's presence, and his willingness to be visited by a group of students.)

Because of the necessity to begin patient care, neither the pre-experience conference nor the head nurse's orientation can be lengthy; both together should probably take no more than 30 minutes. Careful preparation by students prior to their practice period is essential in making effective use of this brief period of orientation to their clinical assignments. The pre-experience conference should focus on the acquisition of knowledge necessary for the immediate care of patients. For example, detailed discussion of the significance of certain tests in arriving at a medical diagnosis is likely to lose the attention of students; but brief, concrete discussion of the symptoms students should watch for tends to hold their interest because it is relevant to their immediate concerns in caring for patients.

TEACHING AND LEARNING IN THE CLINICAL SITUATION

Helping students to adapt behavior to the clinical environment

Clinical teaching has been affected by the fact that, more and more, students move back and forth between campus, home, and clinical environments. Although this change is most pronounced in collegiate programs, it is not confined to them. Students used to live, work, and learn at the hospital. Although they had a great deal to learn about the customs of the hospital community, they had less need to adapt simultaneously to two or three quite dissimilar environments. How different it is now for the young married woman who spends fifteen hours a week in clinical practice, attends classes at a community college, and resumes care of her home and family in the evening! There are many differences in the kinds of requirements made by clinical and classroom environments. Some of the most obvious differences concern dress and etiquette.

After contrasting the casual dress of many co-eds with the neat and conservative dress of most nurses, one may wonder that students manage as well as they do to adapt to such different expectations as those regarding appearance. One sees the student on campus, her long hair falling loosely about her shoulders, shod in well-seasoned sneakers. The next morning, with her hair in an impeccably neat upswept style, and polished oxfords, she is the epitome of the professional student. Her manner has changed, too, from the casual "Hi," to "Good morning, Miss Clark." Not all students, to be sure, make these shifts so successfully.

Recognizing and discussing the differences between behavior expected at the hospital and on campus can help students adapt to them. Too often emphasis is placed upon lists of rules for each setting, rather than upon discussion of the differences between hospital and campus cultures and the possible reasons for these differences; for instance, students are rarely expected to telephone in advance when they will miss college classes or laboratories, but such notification is essential in relation to clinical laboratories. Frank discussion of the reasons for certain rules concerning dress is more likely to gain the students' respect and cooperation than admonitions to "look professional." Differentiating questions of taste from matters of patient safety can assist students in understanding the reasons behind certain rules. Rules against wearing earrings with the uniform have more to do with taste than patients' safety; rules against wearing rings are based on both factors.

A student who is awkward in making the adaptation to differences expected in hospital and clinical settings is occasionally considered to be slipshod or indifferent. Her overly casual manner and dress may be viewed as evidence of general carelessness and lack of aptitude for nursing, while it may actually signify that the student is having difficulty in adjusting her dress and behavior to different environments.

When the students must leave the clinical area, and the patients they have been caring for, to attend classes on a campus which may be some distance from the agency, some vexing problems may arise, necessitating adaptation of practices which are feasible only when students live and attend classes adjacent to the hospital. If the student must catch a bus which runs once an hour, in order to arrive at class on time, she cannot stay on the ward an extra 15 minutes, nor can she dash back between classes with medicine tickets she has found in her pocket. Such concerns seem trivial until one becomes involved in a situation in which tension mounts because the previously employed methods of dealing with a particular situation are not used or, worse still, where failure to use these methods is equated with lack of interest in the patients' welfare. Discussion between agency staff and faculty members when students begin clinical work, concerning ways of handling these situations in light of changed circumstances is essential. It takes extra effort on the part of faculty to see that the number of such incidents is kept to a minimum. Making sure that no student has medicine tickets with her when she leaves the ward, for example, can save time and tempers later, as can an agreement with the head nurse concerning the procedure to be followed if, despite these efforts, medicine cards are removed accidentally. The teacher must convey to students an expectation that they will assume responsibility for completing their assignments and for conscientious reporting to the staff before they leave the unit. Particularly in situations where the instructor is responsible for a small group of students, she must guard against a tendency to shield them from assuming responsibility for their own work, such as finishing the details of charting or making sure that equipment is returned to its place. Students must be helped to gradually develop greater responsibility and independence in relation to their clinical work.

*Helping students differentiate between their needs
and those of their patients*

College students are expected to assume responsibility for themselves and their own learning. Students in the clinical setting, however, are expected not only to assume this same responsibility but, in addition, must necessarily assume responsibility for the welfare of others. (This does not mean that learning and serving are antithetical. Indeed, the essence of nursing lies in learning to serve.) Nursing students are charged with responsibility for others earlier in their post high-school education than are students in most other fields. The difference in amount of emphasis on responsibility for self and responsibility for others is sufficient to cause difficulties for some students and teachers who move between clinical and college environments. In her nursing courses the student is expected to show not only knowledge, but willingness and ability to use the knowledge to serve others. Each time she works with a patient she assumes some responsibility for his safety and welfare. She also is responsible to the teacher, and to the staff of the agency for completing her assignment satisfactorily.

The student who undertakes care of any patient to increase her knowledge is not wholly free to concentrate on her educational needs, nor is the teacher of nursing free to concentrate solely on the educational needs of students. Regardless of the auspices under which the school is conducted, students and faculty must approach all clinical learning situations with a dual viewpoint which embraces both the requirements of the patient and the educational needs of the student. Because such a viewpoint is required, no

matter how inexperienced the student may be, it is not surprising that this is one of the most difficult aspects of clinical
teaching and learning. Helping the student to recognize
nursing needs of patients and to differentiate these from her
own learning needs and interests, when the two are dissimilar, is one of the most challenging dimensions of clinical teaching. The student who is learning to give injections
concentrates on this technique and, until her skill in the
technique has increased, may not recognize the patient's
need for explanation and reassurance concerning the injection. On the other hand, the student who is learning
interviewing skills may be oblivious of the patient's requirements for position change or back care, as she concentrates
on talking with him about his experiences.

Equally challenging is the need for the teacher to be
perceptive not only of how the student is functioning, but
also of the patient's responses. In the concentration on
whether all the assigned medicines are administered correctly one may fail to notice that a patient seems lonely or
discouraged. Unless one is sensitive to the patient, as well
as to the student, one may, without realizing it, expect him
to submerge those needs which do not happen to be convenient at this particular point in the student's learning.
For instance, one may expect him to be pleasant to the student, even though he feels irritable; to comply quickly with
her requests, even though he is hesitant or fearful; to allow
her to care for him, even though he should be helped to
carry out his own physical care.

When the teacher undertakes to provide nursing care for
selected patients as well as clinical experience for students,
it is sometimes difficult not to expect the student to learn
more quickly than she is able to in order to care for the
patient more efficiently. To some extent, this dilemma can
be handled through appropriate clinical assignments (see
pp. 84-95) but, because the situation is seldom such that it is
possible to obtain a perfect match of the patient's needs with
the student's needs, the teacher is, to some degree, always

faced with the problem of satisfying two unlike sets of requirements—those of the patient and those of the student.

Alertness to the needs of the patient, and to what can be expected of the student in a given situation, can help the teacher to proceed in a way which is responsive to the requirements of both patient and student. For example, a freshman was assigned to care for a patient who was recovering smoothly from a dilation and curettage performed for diagnostic purposes. The surgeon arrived, spoke with the patient, and left before the nursing staff or the student had a chance to speak to him. The student found the patient sobbing uncontrollably and, after an unsuccessful effort to find out what was the trouble, sought her teacher's assistance. A glance at the chart before going to the patient's room gave the teacher a clue, for the surgeon had left an order for the patient to be prepared for a hysterectomy. He had also written a "stat" order for a sedative. After alerting the medicine nurse to the order for sedation, the teacher and student both went to the patient. Since the teacher knew that the student had not had enough instruction or experience to enable her to talk with the patient and help her, the teacher did this, with the student quietly observing. Gradually the patient was able to say that the surgeon had told her she must have her uterus out, and that the operation was scheduled for the next day. She became somewhat calmer and began to talk about her dread of the surgery, and of the fact that the news had confirmed her fears that her womb would have to be removed. Then the medicine nurse arrived and gave the injection. The patient soon became very drowsy and teacher and student left her to sleep. In talking about the incident, the teacher helped the student to consider what nursing actions were taken and why, as well as the nursing care which would be required later in light of the scheduled surgery and the patient's response to the prospect of a hysterectomy.

Unfortunately, there are times when the teacher realizes that she has not responded as fully as would be desirable to the requirements of both patient and student in a particular situation. On one such occasion, a student was working with an arthritic patient and, as the morning progressed, the instructor realized that the patient should be encouraged to do more of her own bath, rather than having the student do it for her. The patient moved very slowly, and in order for the bath to be finished in the allotted time, it seemed necessary to have the student continue to bathe the patient. The teacher justified this action to herself on the basis of the student's need for experience with giving a bath, plus the realization that the staff could not take on the patient's bath on short notice if the student did not complete it. But the real reason for the way she had proceeded, she acknowledged to herself, was that she was so concerned with whether the bath was being done correctly, and so busy supervising seven beginners, that it really did not occur to her, initially, that while the student's interests were being well served, the patient's were not. The next day she assigned the same student to the patient, but the assignment did not include making the patient's bed. Thus additional time was allowed to permit the patient to help with her own bath. The teacher explained the desirability of encouraging the patient to do as much as she could for herself, and helped the student determine which aspects of personal hygiene the patient could probably carry out unaided, which she might do with the student's assistance, and which the student would probably need to do for the patient.

There are some situations in which the learning requirements of students are quite frankly uppermost. A patient's cooperation may be requested, for example, in carrying out a demonstration of how to assist patients to get out of bed. Such requests, which are usually made of convalescent patients who seem to enjoy their contact with students and faculty, can be mutually advantageous, providing attention and diversion for the patient and a realistic demonstration for students.

Safeguarding the rights of patients

When asking a patient to participate in any aspect of the teaching program—whether it is to help with a demonstration or to have a student care for him—it is important not to convey the impression that a refusal would be unsporting of him. Despite the requirements of the educational program, it is essential that patients not receive subtle pressure to "cooperate" if to do so would make them uncomfortable. Most of them welcome the opportunity to participate. Advantages to the patient include the opportunity to help with useful work despite illness, and the fact that students can often give them additional time and attention because they are not responsible for the nursing service on the total ward.

Three important measures in safeguarding the rights of patients with whom students work are:

- Asking not only whether the patient wishes to participate, but explaining what such participation entails. A patient may be willing to have one student observe while his colostomy is irrigated, but flinch at the prospect of having three student observers.
- Including him in the conversation; for example, encouraging him to say whether the Ace bandage being applied feels comfortable. Reserving conversation about him and his care, which might be puzzling or worrisome to him, until one is away from him.
- Avoiding any action which would lessen the patient's confidence in the student's ability.

Since it is acknowledged to the patient that the student is learning and that the teacher is there to help her learn, comments made in the patient's presence such as "Let me show you" or "Let's use shorter pieces of tape" or any similar instructions are expected and appropriate. A strained silence and use of gestures and facial expressions to avoid saying anything in front of the patient usually serve only to make everyone acutely uncomfortable. On the other hand, a comment such as "That isn't the way you were taught" or "Wipe from front to back so you don't infect her urethra" can be upsetting by damaging the patient's confidence in the student, or pointing out hazards which the patient may not have thought of.

Perhaps because of the lack of adequate staff in many hospitals, it is easy to underestimate what is being asked of the patient when he has a student caring for him. Particularly at the start of the program, the student lacks the confidence and the sure touch which can help a patient feel more secure. Consequently, the teacher should be present frequently in order to help the patient feel more confident and to observe how he is responding to the care the student is giving. Some patients, although they do not repeatedly ask, "Is she doing it right?" convey this question by a facial expression or by becoming overtalkative or fidgety. They are eager for the teacher to stay with them as much as possible while the student is giving care. In contrast, some patients quickly assume the role of protector of the student. One may walk into the patient's room to find him sitting motionless in his freshly made bed, beaming, and repeating, "She did just fine; everything is just fine." The patient may give the impression that the teacher should not look too closely—that he will protect the student not only from criticism but also from the supervision which might lead to it. Some flexibility of approach to supervision may be helpful in this instance. One can, for example, have more discussion with the student while she is away from the patient and thus cut down on the amount of supervision which takes place in front of him. It may be necessary, too, to explain the purposes of supervision more fully to this patient and point out to him that the process is a way of teaching rather than an attempt to get the student into trouble. If a patient shows by his actions (seeming very apprehensive, for instance) that having a student care for him is upsetting to him, he should not be assigned to a student again, unless he and the teacher can discuss the problem and he can be helped to feel comfortable about receiving care from a student.

Most patients can take in stride the fluctuations in the amount of nursing time spent with them by students and registered nurses, since they recognize that it usually takes a learner longer to carry out an activity than it does an experienced person, and also that students usually do not have responsibility to see that all patients on the ward receive care. An explanation of these differences will be helpful to those patients who find this fluctuation in the amount of time spent with them puzzling.

Fluctuations in the plan of care are another matter, however. Differences in the way his care is to be approached should be ironed out by staff, students, and faculty, thus avoiding subjecting the patient to confusing and possibly harmful differences in techniques. If, for example, the patient says to the student who is changing his dressing, "That is not the way the nurse did it yesterday," the student should be encouraged to ask the patient to explain the difference (rather than using a comment like, "Well, all nurses have their own ways of doing things, I guess"). If the difference is merely a matter of preference, the student can explain this, and perhaps use whichever method the patient is accustomed to. On the other hand, if the difference is one which can affect the outcome of the treatment, the student should ask the teacher which method is to be used before proceeding further.

Student response to patients

A great deal has been written concerning the student's response to the patient. Nevertheless, a few aspects of this topic which sometimes present difficulties in the clinical setting will be mentioned here.

Increasing numbers of patients are in the older age groups, and consequently students are likely to have considerable contact with these patients in a variety of settings, such as nursing homes, hospitals, and the patients' own homes. Sometimes students find it especially difficult to encourage an older person to follow a part of his treatment which he finds unpleasant. Although the teacher and student talk together about the patient's requirement to be helped to get out of bed, the patient may remain in bed. The procedure may be quite familiar to the student and seem to present no special difficulties. Why, then, does the patient continue to lie in bed? Observation of the way the student and patient relate to one another may disclose that they are behaving more like grandfather and granddaughter than nurse and patient. Her youth and lack of assurance may lead him to give the directions, and her to follow them. The patient finds all sorts of reasons why he should not get up, and the student acquiesces. The teacher's intervention in such a situation can seem unwelcome to both of them. After all, were they not getting along fine? Raising the question of getting up seems like rocking the boat. It is essential to help the student understand, as fully as possible, why the patient is to get out of bed, and to recognize that nursing involves not only doing what patients request but also helping them to accept treatments which they may prefer to omit. It is important to remember, when explaining to the student why the patient should get up, that her frame of reference is different from one's own. To her way of thinking, the hazards of prolonged bed rest for an eighty-year-old man may seem a trifle exaggerated. If she has never seen a decubitus ulcer or been astonished at the rapidity of its appearance, or observed the swiftness with which an older person's shallow breathing and immobility can give way to pneumonia, the possibility of these complications may seem remote. Acknowledgment of this difference in attitude, and discussion of probable reasons for it, can help teacher and student to proceed, if not in complete agreement, at least with recognition of some of the differences in their points of view. One aspect of being a student involves following the recommendation of a more experienced person, even though she may not, at the time, fully appreciate the importance of the particular recommendation. The teacher may decide that she herself should talk with and assist the patient out of bed (aided by the student) since the student is not handling this aspect of care satisfactorily, and then assign the student to do it alone the next morning.

A similar problem may arise when students are asked to carry out painful or uncomfortable treatments for a patient of any age. Perhaps the patient does not do his exercises, although there seems to be time enough; perhaps one who finds turning painful is always on his back. It is important for the teacher to be alert to these matters, so that they are not casually ascribed to lack of time. The student may not realize that she is, herself, avoiding some necessary aspects of care. Encouraging her to discuss the difficulties she is having with turning a reluctant patient, assisting her with the procedure at first and, if necessary, giving further explanation of the reason for the procedure can help the student carry it out more independently the next time.

Participating with students in direct patient care

The teacher should not only supervise the care which students give, but give direct patient care herself in the presence of one or several students. The value of observing the teacher's work with patients is sometimes overlooked. The bed bath may be demonstrated in a classroom; subsequently the teacher observes the students' bathing of patients. In this instance, students do not have opportunity to observe the teacher giving a bath to a patient (or performing whatever aspect of care is being taught). By observing the teacher's care of patients students observe not only the procedure, but the interaction between the teacher and the patient, as well as the adaptations in care which are necessary due to the patient's condition or to the particular hospital environment. They are also helped to gain confidence that the care being demonstrated is really possible to perform. Thus, participation in patient care is necessary in order to provide a role model for students (see Chapter

3), but an equally important result is that it allows the teacher to gain information about the patient which enables her to give more meaningful instruction. First-hand information permits her to validate with the student such questions as: How much drainage was on the dressing? Did it indicate fresh bleeding? Did the patient seem discouraged? The teacher can also bring to the student's attention such an additional aspect of the situation as the fact that the patient's wife is pacing up and down in the solarium, and that no one has spoken with her since her arrival on the ward.

It is not possible for any teacher to participate in, or even observe all nursing care given by her students. Some of the discussion which occurs with students individually and in groups is, therefore, based wholly on the student's recall and reporting of the situation. However, the more opportunity the teacher has to observe directly and to participate in the patient's care, the better prepared she will be to discuss his nursing requirements. Television is useful in extending the range of the teacher's observations of students' clinical work, as well as increasing their opportunities to observe the teacher's care of patients.

Channeling student questions

Beginning nursing students usually have little awareness of when they need the teacher's help and when they can proceed without it. But as they gain in experience their judgment concerning when they need help, and how quickly they need it, improves. When teaching a group of students during their first months of clinical practice, the teacher may have several students waiting to ask questions at the same time. As she listens to each, she decides which student needs assistance most promptly. Perhaps one of the questions can best be answered by the ward clerk, and the student is referred to her. More advanced students can channel many of their questions to various other personnel. They do a good deal of priority-placing themselves, waiting until an opportune moment to ask questions which are not pressing.

It is important for the teacher to consider what kinds of help she can and should give students, and what kinds they should seek from others. (See Chapter 3.) Students and faculty may function as a group apart, having little inter-action with ward staff. This can be a distinct disadvantage to students in their later practice, because they have not had opportunities to learn gradually about the roles of staff members and how to work with them. For example, if the student requires a strong arm to help her lift a patient, but knows how to do it, it is more appropriate for her to ask the orderly's assistance than to seek the teacher's help.

The teacher must be alert to the numbers of questions and requests students are making, so that certain members of the staff are not interrupted unnecessarily. One means of cutting down the number of separate questions being chan-neled to the head nurse, for instance, is to plan with her for an effective orientation to the ward, and to notice the kinds of questions being raised once students begin their practice. Perhaps some of these queries can be handled by group discussion and demonstration after the first few days of practice on the ward; some may also be included in the orientation period for the next group of students.

What about students asking each other for help? One of the first and most natural ways in which students learn to work with others is to begin doing it within their own group. The teacher can suggest, when the morning assign-ment is given, that one student is likely to need assistance, and suggest a student who may be free at the time to give it. Such arrangements often assist the helping student to gain necessary experience; while assisting her classmate to make an occupied bed, she may have her first opportunity to observe a spica cast. Students need guidance in learning to differentiate between the type of help they may seek from classmates and the type they should seek from the instructor. Occasionally, a student will make every effort to avoid asking the teacher a question; instead she attempts to get the information from a classmate. In such an instance, a discussion with the student concerning her reluctance to seek the teacher's help is necessary. On the other hand, the sharing of some kinds of information among students is natural and sensible. Realizing that one student has shown another where the urinals are kept, or how to work a high-low bed should not make the teacher feel that her authority is being threatened.

Dividing instructional time among students

Since individual clinical teaching is so important a part of the educational process, the way the teacher divides her time among students is important. Each student has equal call upon one's time and effort, although the kind of assistance necessary and the timing of it will be different in each instance. On a day when Miss Brown has an assignment which is very new to her, it is natural that she will require more of the teacher's assistance than Miss Evans, who is caring for the same patient she had on the previous day and is oriented to many of his nursing requirements. When reflecting upon a week's or a month's clinical teaching, one should be able to trace clearly a pattern of instruction with each student. Sometimes, though, upon looking back one notes that the performance of one or two students seems to fade into the shadows. Comments on anecdotal records have a vague quality, and one is left with the hazy impression that the student performed adequately. But the files are full to overflowing with notes about Mary Jane Doe who, it seems, has required almost constant supervision lest someone be harmed by her frequent lapses of judgment. The dilemma of providing adequate supervision for a slow student while at the same time teaching the rest of the group is a familiar one. Often it is the brighter, more capable student who is short-changed in this situation, because one can be reasonably certain that she will perform adequately with a minimum of help. This student, though, requires challenge and teaching which helps her use her abilities more fully. Some setting of limits is necessary in these situations. A difficult but important aspect of the teacher's responsibility involves considering how long a very slow student can continue to receive more instruction than the others without jeopardizing their learning. If a point is reached where the teacher must consistently devote so much time to one student that instruction of the others is compromised, she has the obligation to state to the Dean, or perhaps to a faculty committee, that the needs of this particular student for supervision exceed what can be provided in the particular situation.

Helping students to handle the multidimensional clinical situation

Even after apportioning her time in what seems to be a reasonable way, the teacher may find herself thinking that she has done a lot of dropping in and out of clinical situations, but is not sure what has been accomplished. Clarity about what one plans to observe is essential; however, the teacher should not be so engrossed in following this plan that she fails to notice unusual or unexpected aspects of the situation. The purpose of a visit may be to observe how effectively a student is demonstrating the bathing of a newborn. However, if the new mother has severe varicose veins, one should also note that, although the baby's bath is being demonstrated skillfully, the mother is standing during the demonstration, rather than sitting with her legs elevated.

Sometimes, in the process of going from student to student one makes another kind of unanticipated discovery. The student has gone ahead with a procedure when she should have waited! Particularly during the period when teacher and student are not accustomed to working together, it is important to tell the student very specifically when she is expected to wait until the teacher arrives and when she may proceed without the teacher's presence. Sometimes the teacher is so certain of her own intentions that she assumes that the student also knows what they are. Although the clinical instructor is present during parts of all of the students' various activities, deciding what parts she should observe during each practice session requires some thought, not only to prevent her from repeatedly observing the same student performing the same skill, but also so that she does not neglect to consider her legal and ethical responsibilities. If the student requires direct supervision with the preparation of medications, for example, the teacher should be present and should observe whether the correct medicine is prepared in the correct dosage. Looking at a tray of medicines after they have been poured is not effective supervision and merely takes time which could be devoted to other matters; it may also give both teacher and student a false sense of security by leading both to think that the medicines have been checked when, in fact, they have not.

Often one assignment can fulfill several purposes. It is not unusual for a student to under-utilize an assignment by concentrating on one aspect of it and neglecting others, even when time permits fuller utilization. The student who is "all finished" now that she has completed the patient's morning care is a familiar example. She may be ready to assist her classmates when the teacher points out the other aspects of the learning experience which had escaped the student's notice. Perhaps this is a time when the student can arrange to talk with the patient's wife about the dressings her husband will need when he goes home, and begin teaching her how to apply them.

Although students cannot be expected to recognize the multiple factors in a situation all at once, or without guidance, they can, as they have repeated experiences, be helped to broaden their understanding of the many facets of the patient's situation. For example, the student who goes weekly for one semester to a nursing home to interview one of its residents may, as the weeks progress, be helped to appreciate some of the differences and similarities of this setting as compared with a general hospital and also to consider some ways in which these differences affect the nurse-patient relationship. In addition, her attention and study may be directed toward some typical problems and concerns of aged people who are in nursing homes, thereby sharpening her awareness and her interest in the way her particular patient's concerns are similar to or different from those she has read about. Greater continuity of learning and more meaningful learning can occur by using each experience more fully, rather than by having a great variety of experiences which are incompletely or superficially utilized.

Helping students to get the most
from observational experiences

Now that students are not relied upon for providing nursing service to the extent they were in the past, there is greater opportunity to broaden their learning by including such observational experiences as visits to well-baby clinics, schools, day centers for healthy older people, nursing homes, and sheltered workshops. Sometimes such experiences are over-used during a burst of enthusiasm for their merit. Increasing greatly the number of observational experiences while decreasing the opportunities for students to practice, if carried far enough, can interfere with their acquisition of skill in giving care and can interrupt the continuity of their practice. When wisely chosen, however, they enrich the educational program tremendously. How much more valuable it is for students to actually see an industrial health center, and to talk with nurses who work there, than merely to read about the existence of such facilities! By such direct observation the student can learn a great deal concerning the differences and similarities in the nurse's role in various settings, and can develop a more realistic view of the way in which she can work with staff in various types of health agencies.

What measures can add to the usefulness of observational experiences? It is common for a teacher to draw a student aside and say, "Come, look at this, and we'll talk about it afterward." Valuable opportunities for observing often arise very unexpectedly, and the only way to utilize them is to have the student observe on the spur of the moment. In such instances it is essential not to omit the discussion later; otherwise what the student has seen may serve more to confuse than to teach her. Whenever possible, however, it is desirable to have some orientation, beforehand, to the purpose of the observation and to some of the main aspects which the student is expected to notice. This orientation and the subsequent discussion after the observation should not be so rigidly structured, though, that the student is discouraged from making other observations and sharing them with her classmates. During a discussion of a visit to a sheltered workshop, for instance, a student may voice astonishment that nursing was actually being carried out by people wearing street clothes—a point which may have seemed so commonplace to the teacher that she had not thought of bringing it up for discussion. Sharing the actual observation can be advantageous to the teacher, because it gives her an opportunity to contrast her own impressions with those of students and also to pass on to them certain factual information which she has acquired through experience; it also makes it easier for her to note and correct misinterpretations which students sometimes make due to their lack of knowledge or experience, or simply because they did not have a chance to observe the entire situation. A student who has observed the passing of a naso-gastric tube may say, "But neither the doctor nor the nurse explained anything to the patient. They just pushed the tube down." Such a comment is understandable if the student has arrived just as the doctor started to pass the tube. The teacher may mention that an explanation was given to the patient before the procedure was begun. Perhaps further discussion indicates that the student thought the patient received "no reassurance" during the procedure and, as the conversation continues, it becomes clear that she is referring to verbal reassurance. The teacher may then ask what other ways there are of reassuring patients during uncomfortable procedures besides talking to them. The student may respond by suggesting the use of measures to help the patient cope with such embarrassing reactions as coughing, spitting, or regurgitating. Attention could be drawn to the fact that the physician and patient communicated by nods and facial expression, and that the doctor did not rush the patient, but carried out each step of the procedure when the patient indicated he was ready to continue. If possible, the discussion of what was observed should take place promptly, so that the student can draw upon still-vivid impressions.

THE POST-EXPERIENCE CLINICAL CONFERENCE

The teacher who recognizes the value of clinical conferences in extending the range of students' learning is less tempted to feel that every student must be provided with a great variety of experiences. Some teachers are vulnerable to this pressure, despite their recognition that no program can provide a sampling of all the kinds of situations a nurse is likely to meet, and that, even if it could, such an array of experiences would afford little opportunity for consecutive experience with any one patient.

Post-experience clinical conference discussions are usually more detailed than those of the pre-experience conference. Ideally, each student should be given time to discuss her observations and the care she gave, and to receive suggestions from her teacher and classmates. In some programs it is not possible to attain this ideal, because of the shortage of faculty members to lead conference groups, or perhaps because of the difficulty of providing both liberal-arts and professional education in the limited time available in an undergraduate program. Until these problems are solved, the teacher will need to make adaptations in the way conferences are conducted. In order to permit thorough discussion it may be necessary to rotate opportunities for students to present their experiences, depending to some extent on the size of the group and the length of time available for the conference. If, for instance, 12 or 15 students are asked to discuss their experiences during a 1-hour conference, the content may never move beyond the most superficial and general aspects of each situation. Students may mention the need for reassuring patients' families, but somehow there is never time for consideration of how the nurse can go about providing such reassurance. If it is not possible to have a smaller group, or a longer conference period, having half of the students present material at one conference and half at the next will permit fuller discussion of each student's data.

Another value of rotating opportunities for students to discuss their experiences is that it provides a more equal chance for the shy and the talkative students to present material. Since the experience of discussing data is especially helpful to the student who presents it, it is essential that all students in the group have equal opportunity to do so.

Helping students to prepare for the conference

How well the student is prepared for the post-experience conference depends, to a great extent, on how well she has prepared for her clinical assignment. She should know at least one day in advance who her patient or patients will be. The degree to which she is expected to use her own initative in preparing for her clinical and conference participation is closely related to her level of learning. The teacher may give a freshman such directions as these: "Find out about the usual symptoms and treatment of the illness. Look up the action of each drug your patient is receiving. Be ready to describe the care you gave to the patient and your reasons for proceeding as you did." The more advanced student is expected to show greater ability to decide what is important to prepare, and how to go about doing it. For instance, the teacher may ask her to develop her own observation guide in preparation for a home visit, and to discuss this guide with the teacher before making the visit. Students who have prepared thoughtfully can use their practice periods to greater advantage; they also come to conference with more to share with their classmates. In addition, less conference time is wasted in such activities as drilling students on drug actions and definitions, which penalize particularly the more responsible and able students who soon lose interest if this becomes the focus of the conference.

The amount of time scheduled for clinical practice and the amount allotted for the post-experience conference have a reciprocal relationship. If the conference period is so lengthy that practice is curtailed, students do not have a chance to develop the necessary skills or to become well enough acquainted with the patient's care to discuss it intelligently. On the other hand, if the conference period is too brief, students do not have sufficient opportunity to analyze and learn from the experiences they are having. The time allotment for practice and conference depends on the purposes of the course. If the purpose of a senior course is to help students get ready for the role of staff nurses and to learn to deal more capably with the stresses involved in providing nursing care in a busy agency, a considerable portion of their time should be spent in practice. A ratio of two hours of practice to two hours of conference would, in this case, be less desirable than, say, six hours of practice and two hours of conference.

Sometimes the problem lies not so much in allotting sufficient time for conferences as in seeing that all students arrive with reasonable promptness. There are occasions when students are unavoidably delayed, and a teacher should not imply that unforeseen but essential needs of patients should be disregarded so that students invariably arrive on time. This is different, though, from conveying the attitude that the conference is less important than practice, and that repeatedly missing half of a conference is acceptable. Possibly students are late because their assignments are too heavy, or because they need to learn to organize their activities better. Maybe it is because no satisfactory plan has been worked out with the staff about taking over the care of assigned patients when students are expected to leave the ward. Considering and dealing with the cause of lateness is important, not only in conveying the expectation that all students will attend the entire conference, but also in showing willingness to work with students in solving problems which may be interfering with their attendance.

Combining demonstration with the conference

Combining a demonstration and conference is useful in many situations in which the opportunity to observe an aspect of care can make subsequent discussion more vivid and more relevant to the real situation. Students' reactions to some nursing situations are sharpened and their awareness is increased more by direct observation than by reading and discussion alone. Observation also helps them to develop a more realistic appreciation of what the situation may mean to the patient. While observing the suctioning of a tracheostomy and the cleansing of the inner tube, for example, a student may feel uneasy because she fears that the patient may not be able to breathe through the tube; she may also come to appreciate more fully the apprehension of the patient whose breathing depends upon prompt suctioning of a tiny tube which is often partially plugged with mucus. During the conference which follows the demonstration, she can bring up specific questions which might not have occurred to her if she had not observed the procedure.

This kind of demonstration is an effective way of sharing available experiences without resorting to such frequent changes in assignment that continuity of experience is lost. Rather than rotating the experience with one tracheostomy patient so that each student in the group has the patient one day, it may be more effective to plan consecutive assignment of one or two students to care of this patient, and share some aspects of the care with the rest of the group by means of demonstrations. In this way the students who are assigned to the patient's care can present more complete and useful discussions during the conference, since they necessarily become involved in many aspects of the nursing situation.

Demonstrating some particular aspects of nursing care to groups of students is a more effective use of the teacher's time than showing each student individually how to carry out a special type of care. Sometimes teachers are reluctant to plan such demonstrations lest they be considered "procedure-oriented." Whether one's teaching is oriented broadly to many nursing requirements or narrowly to procedures depends upon the beliefs of the teacher as these are reflected in the way she handles each situation. One teacher can plan a combined demonstration and conference which deals not only with the procedures of changing dressings and doing irrigations for a patient with a colostomy, but which also includes such considerations as helping the patient come to terms with the fact that he has a stoma, and with the changes in his life which are necessary because of it. Another teacher may emphasize the care of the stoma almost exclusively. Carrying out the demonstration for each student individually is no guarantee against overemphasis on procedure; one can be as "procedure-oriented" in teaching one student as in teaching five.

If the teacher decides that a demonstration is necessary and worthwhile, she should convey this conviction in the way the demonstration is carried out. Suppose, for instance, that the teacher is demonstrating the giving and removing of bedpans and urinals. It is important for her to discuss such matters as students' feelings about caring for the excreta of others, and some possible reactions of the patient to requiring this assistance from another person. The demonstration itself is also important, and sufficient time and thought should be given to it to carry it out effectively. The lack of sureness about one's convictions concerning the usefulness of certain learning experiences can lead to teaching which is carried out hurriedly and apologetically.

Sharing demonstrations encourages students to validate with each other what has been taught; for instance, after several students have observed the removal of the inner tracheostomy tube for cleaning, one of them may be confused about whether it was the outer or the inner tube which was removed. Classmates who shared the experience can recall that it was the inner tube.

Although there are many acceptable ways of performing almost any nursing action, it is helpful to students to have the teacher who gives the initial demonstration stress one acceptable way, while indicating that there are other methods which students will see used, and that they themselves may use later. In her concern not to present material in a cut-and-dried fashion, such as "At this hospital we fold the linen in quarters as we remove it from the bed," the teacher may present so many alternatives that students are bewildered, or she may present the material so vaguely that it is hard for the student to proceed when she encounters the actual nursing situation. Saying, "The important point about removing bed linen is to do it neatly," then deftly folding the linen without making any comment about *how* one is doing it, can make this particular act seem very simple to the student while she is observing, but she may become tangled in the sheets when she herself tries to do it. Lack of definiteness about the way the student is expected to proceed can result in spending more, rather than less, time on something which is simple and should be handled with dispatch so that the student can go on to learn about other aspects of care. When in a new situation, the student needs the security of knowing how to carry out certain manual skills which in themselves may not be of great importance, but which, if handled smoothly, can help her to remain poised and to become more confident as she works with patients.

Judgment is necessary concerning which topics require

discussion. Is it a worthwhile use of time for students to discuss the possible ways of folding a sheet, and the merits of each, in light of the many other topics which require discussion? The initial demonstration of one method does not necessarily conflict with fostering flexibility in learning different methods. Encouragement of flexibility can occur promptly and deliberately, as soon as students have gained some experience and confidence with the first method they learned. "Why not try it this way this time?" is a good question for the student who is ready for it, and can help her to become more flexible in the way she approaches patient care.

Usually fewer students can be accommodated during the demonstration than during the conference. Thus, with the patient's permission, two groups, each comprising four students, may observe a demonstration, and all eight students attend the conference. Much depends upon the patient's reaction to the number who observe, the size of the room, and how well students can see under the particular circumstances. Films and closed-circuit television are effective for certain demonstrations. They have the advantage of accommodating larger groups of students and of protecting patients from having groups observe such intimate procedures as perineal care or the expression of milk from the breast.

SUGGESTED READING

Ackert, Helen O. "A Tailor-made Caseload for Students," *Nursing Outlook, 13*:46, April, 1965.

Burrill, Marjorie. "Helping Students Identify and Solve Patients' Problems," *Nursing Outlook, 14*:46, February, 1966.

Clissold, Grace K. *How to Function Effectively as a Teacher in the Clinical Area.* New York, Springer, 1962.

Coston, Harriet M. "Patient Centered Teaching," *Nursing Outlook, 6*:697, December, 1958.

Faddis, Margene O. "On Clinical Teaching," *American Journal of Nursing, 60*:1461, October, 1960.

Griffin, Gerald J., Kinsinger, Robert E., and Pitman, Avis J. *Clinical Nursing Instruction by Television.* Philadelphia, J. B. Lippincott, 1965.

Hassenplug, Lulu Wolf. "The Good Teacher," *Nursing Outlook, 13*:24, October, 1965.

Hayter, Jean. "Guidelines for Selecting Learning Experiences," *Nursing Outlook, 15*:63, December, 1967.

Heidgerken, Loretta E. *Teaching and Learning in Schools of Nursing.* 3rd ed. Philadelphia, J. B. Lippincott, 1965.

Hill, Richard J. "The Right to Fail," *Nursing Outlook, 13*:38, April, 1965.

Ingles, Thelma. "On Developing Skilled Practitioners," *American Journal of Nursing, 60*:1482, October, 1960.

Johnson, Betty Sue. "The Meaning of Touch in Nursing," *Nursing Outlook, 13*:59, February, 1965.

Lefkowitz, Annette S., and Hart, Ann. "Selection of Clinical Laboratory Facilities for Collegiate Nursing Education," *Nursing Forum, 4*:95, No. 1, November, 1965.

Major, Dorothy. "Keys to a Philosophy of Teaching," *Nursing Outlook, 10*:510, August, 1962.

Martin, Almeda Biggs. "On a Patient-Centered Approach," *American Journal of Nursing, 60*:1472, October, 1960.

Moore, Lucille, and White, George D. "Comparisons of Teaching Methods in Maternal and Infant Care," *Nursing Outlook, 13*:74, May, 1965.

Mullins, Agnes P. "First Clinical Assignments," *Nursing Outlook, 13*:47, February, 1965.

Nuckolls, Katherine B. "Tenderness and Technique via T.V.," *American Journal of Nursing, 66*:2690, December, 1966.

Quint, Jeanne C. "The Hidden Hazards in Patient Assignments," *Nursing Outlook, 13*:50, November, 1965.

Quint, Jeanne C. "Hidden Hazards for Nurse Teachers," *Nursing Outlook, 15*:34, April, 1967.

Schumann, Delores M. "An Improved Method of Making Clinical Assignments," *Nursing Outlook, 15*:52, April, 1967.

Smith, Dorothy W., and Gips, Claudia D. "Medical-Surgical Nursing in an Associate Degree Program," *Nursing Outlook, 4*:349, June, 1956.

Ujhely, Gertrud B. *Determinants of the Nurse-Patient Relationship.* New York, Springer, 1968.

6

Helping Students to Develop Independence

Helping students gradually to develop greater independence in their clinical practice is a challenge which has many different aspects. To practice independently means that one is able to function in nursing situations as they exist today. How can the teacher help the student to acquire the skills and abilities which will be expected of her as a practitioner in current nursing situations? How can the teacher encourage thoroughness and pride in work well done, without allowing the emphasis upon completion of each detail to interfere with instruction about larger issues and concepts? What approach is practical and realistic enough in its orientation to enable students to survive professionally under the hectic conditions so prevalent in nursing today, yet has sufficient vision and idealism to encourage students to work toward changes in the way nursing care is given?

Some approaches which help the student develop independence include: fostering responsibility and thoroughness; gradually increasing the complexity of assignments; helping students learn some ways of dealing with stress; fostering mature behavior; and helping students deal with the discrepancy between ideal and actual patient care.

Fostering responsibility and habits of thoroughness

In years past (and even today in some nursing schools)

the student's sense of responsibility and her habits of thoroughness were fostered largely in the context of her role as junior staff member. When she assumed charge of a ward at night she was held accountable by the head nurse and supervisor for carrying out all essential details thoroughly and correctly, just as would be expected of any staff nurse assuming charge of that unit. But now that greater emphasis is placed upon students' learning, and teachers have assumed more responsibility for clinical as well as classroom instruction, the job of seeing that the student completes necessary details falls to instructors. The context is no longer that of head nurse to junior staff member, but that of teacher to student. The teacher is, appropriately, oriented more toward seeing that the student develops the necessary knowledge and skill, and less toward seeing that ward tasks are completed. In her desire not to allow the emphasis on well-scrubbed bath basins and neatly made beds to obscure other aspects of learning, the instructor may pay too little attention to these details. In her concern about providing her students with as many and varied experiences as possible she may give students larger assignments than they can complete. While failure to complete an assignment is sometimes unavoidable, its frequent occurrence can interfere with the student's learning to assume responsibility for her own work.

Clarity concerning what should be assigned in order to provide required learning, and planning assignments so that they can be completed in the allotted time are essential.

When a student is learning to take and record temperature, pulse and respirations, for instance, the charting of her findings is an essential part of the experience, and each student should be expected to complete it thoroughly and correctly. On the other hand, once students know how to chart T.P.R.'s there is no reason why the ward clerk should not chart them for students' patients, just as she does for other patients on the unit. Therefore students are not assigned to chart T.P.R.'s at that stage of their learning. This is quite different from assigning students to take and record T.P.R.'s and, when a student does not complete the charting, saying, "All right, as long as they're in the T.P.R. book, the clerk will chart them." If a student needs experience in teaching a diabetic how to give himself insulin, and time does not permit her also to bathe him and make his bed, she should not be assigned to do these things, because she is expected to carry out thoroughly her primary assignment which is the teaching.

In the educational context of clinical practice, thoroughness is best taught, not by insistence upon performance of repetitive tasks in order to get the work done, but by expecting each student to complete all aspects of an assignment which has been planned to provide the learning which she requires. While in previous years some demands made upon students for thoroughness and discipline constituted exploitation of the student, today's students must be helped to develop pride in work well done and a sense of responsibility, while obtaining varied and stimulating learning experiences.

Gradually increasing the complexity of assignments

As the student progresses in the program she requires opportunities for experiences different from those she had as a freshman and opportunities to care for patients with more difficult nursing problems. Such experiences used to come to students as a result of nursing service requirements; now they must be deliberately sought.

It is also necessary for the teacher to provide the student with a definite progression in her clinical assignments, so that not only does she refine the skills which were introduced at the beginning of her program, but she also adds new dimensions to her clinical practice. Although new dimensions of nursing practice may be discussed in class, the student often does not have sufficient opportunity to practice them clinically. For example, as we note the way a freshman and a senior student go about their clinical assignments,

are there clear-cut differences in the purposes of their assignments which are apparent to both students and teachers, or
are both students giving the patient's bath, making his bed,
and checking his vital signs, with the main difference being
that the senior does so with more speed and confidence?
Although students need to repeat experiences in order to
enhance certain skills, they also require opportunities to
practice different dimensions of nursing as their program
unfolds. This progression should involve more than learning
certain adaptations of care necessary with different categories of patients and in different settings, although such
learning is part of the progression of the student's education.
Not only, for example, should the student learn how to
adapt the bath procedure from that required for an adult
to that for an infant, or what special precautions are necessary when giving medicines on a psychiatric unit as compared with those required on a medical-surgical unit; she
must also add such further dimensions to her practice as
experiences in teaching groups of patients, teaching patients' families, providing leadership for improvement of care
by nonprofessionals, assuming responsibility for care of
groups of patients, and working with critically ill patients.
    Because we are deeply concerned about protecting the
student from exploitation, we sometimes place so little emphasis upon care of groups of patients and critically ill
patients that her transition to the role of staff nurse is unnecessarily difficult. The degree to which we should emphasize preparing the student for the current realities of
nursing, and the degree to which we should emphasize preparing her for her future role as a professional nurse are
difficult to determine. Although we must strive to prepare
enough professional nurses to give individualized care to
patients, it is nevertheless true that after the student graduates she will work in situations in which she must plan the
nursing care for groups of patients, and in which the amount
of time that can be spent with each patient is sharply limited. Thus, not only does the student need opportunities to

concentrate on care of individual patients; as she progresses in the program she also requires experiences in working with groups of patients in order to develop her ability to organize their care, to learn to recognize which nursing needs require her attention the most, and to develop skill and judgment in delegating certain aspects of care to others.

An important result of these experiences is that they prepare the undergraduate to assume some leadership responsibilities in working with nonprofessionals. On the one hand it seems incongruous to expect any student to prepare to assist others to improve patient care since she, herself, is but a beginner in all areas of nursing. How can she be expected to guide others when she is in the process of getting her own bearings? It would be sounder educationally to place all preparation for leadership at the graduate level, but this approach does not take account of the reality of the current nursing situation, where those who have had only undergraduate education are expected to provide guidance for nonprofessional workers. It seems essential, then, to help prepare the undergraduate for leadership by making sure that she has such experiences as conducting small group conferences, orienting others to nursing needs of patients, and assigning others to various aspects of patient care.

Toward the end of her program, some experience which requires the student to handle larger assignments and to function more independently in relationships with staff, can help her to bridge the gap between what is expected of her as a student and what is expected of the professional nurse. Of course what is expected of nurses is often highly unrealistic (such as assuming responsibility for care of fifty very sick patients), and the concern should be rather to modify the expectations than to attempt to prepare the student to give care under such conditions. Nevertheless, the experience of planning and carrying out care of perhaps ten to fifteen patients, with assistance of auxiliary personnel, is essential for the student whose previous assignments have not exceeded two or three patients, and which have emphasized the nurse's role in giving complete care. Attention can be directed toward developing skill in planning and organizing care of groups of patients, and toward consideration of the student's interaction with staff, and the ways that this affects her ability to provide care. Emphasis should be given to the leadership role of the professional nurse in assigning and guiding auxiliary nursing personnel in aspects of patient care which are appropriate to their level of education and experience. Seminars during which the student discusses her experiences can help her learn how to orient an auxiliary worker to an assignment, how to determine priorities of nursing needs, and so on.

During this period of her program, particularly, the student should be encouraged to consider the effects of working under tension, and some ways in which the undesirable effects of tension can be mitigated by the nurse. Unless opportunities are provided for the student to work in stressful situations and to develop her ability to tolerate and to learn ways of lessening the strain, her ability to use her clinical skills will be lessened. Probably the most important prerequisites for dealing with stress are possession of the knowledge and skill demanded by the situation, and confidence. The student's confidence in her ability is built on mastery of various aspects of nursing practice. As the number of hours allotted to clinical laboratory decreases, it is especially important to arrange the student's schedule so that she spends enough consecutive time with patients to be able to implement some of her plans for nursing care and to evaluate their effectiveness. If the schedule permits only brief patient contact in a situation where there is high turnover of patients, she will have many opportunities to learn to recognize nursing needs, but there will be few occasions when she can follow through with a plan of care. Under such circumstances, the student's confidence in her ability to give care is usually lacking. Of course, the extent to which the student can develop skill during an undergraduate program is limited. Much of this must and should occur after graduation. However, the pendulum sometimes swings so far away from nursing's traditional emphasis upon practice in order to acquire skill that some graduates are unduly hampered by lack of confidence in their abilities as practitioners.

In addition to helping the student acquire knowledge and skill, and fostering the development of confidence which comes with successful performance, the teacher can help her to identify certain strategies for dealing with stress, and to learn to use them knowingly. For instance, one difference between an experienced practitioner and a neophyte is that the former is better able to concentrate on the task at hand. Her skill and knowledge quickly swing into action in response to a demanding situation, facilitating her ability to focus on it. The inexperienced person, in contrast, is more likely to absorb and react to many facets of a tense situation without differentiating which aspects require her attention. Because she has less skill and less ability to discriminate between the more and the less essential elements, she is likely to be overwhelmed by a stressful situation. The student can be helped to recognize the importance of concentrating on what she is doing, determining the relative importance of the various aspects of a situation, and continuing to improve her clinical skills so that she can function adequately under stress. Another important consideration in helping the student deal with stress involves assisting her to recognize the importance of concentrating initially on concrete aspects of care, particularly in situations which arouse her anxiety. The experienced nurse, for example, knows the value of concentrating on the concrete problem of getting an accurate blood pressure reading of a patient in shock, instead of letting her mind race frantically over the various indications of how sick the patient is, and the possible reasons for it. Really concentrating on saying a few words clearly and slowly to a confused patient, thus helping him orient himself to his surroundings, can bring a disturbing and possibly dangerous situation back into control.

The discrepancy between the maturity required of a nurse and that possessed by most undergraduate students presents a dilemma for the teacher who is trying to help students develop greater independence and responsibility. Because nursing students are expected to function clinically at a younger age than most other members of the helping professions, it is inevitable that they and their teachers will encounter considerable strain. The teacher may catch herself becoming impatient with the way a student is handling a situation, and then realize that she is really wishing the student were not behaving like an eighteen-year-old! The giggling at the nurses' station, the awkwardness in approaching a physician, the ill-timed youthful candor which occasionally sparks an explosion in staff-student relationships are all examples of situations which can lead the teacher to wish her students could acquire—in the space of a few days—ten more years' experience in living.

The teacher can foster the student's maturity by consistently orienting herself toward the more adult aspects of her behavior. Treating her as a junior colleague, rather than as a juvenile, but doing so with a light touch which does not squelch her youthful exuberance, is one way to help the student learn to recognize the behavior which is appropriate in different settings and, when necessary, to remind her in a manner that does not humiliate her, that it is time to recapture her professional poise. A look or a humorous comment from the teacher may be all that is necessary to remind students who may be returning from the coffee shop to lower their voices.

Helping students deal with the discrepancy between ideal and actual patient care

Regardless of the setting in which she teaches, the clinical instructor is repeatedly confronted with the problem of how to help students cope with the discrepancy between available knowledge and its application to the care of patients—between what nursing is and what it should be. The gap between the ideal and the reality is sometimes so wide that students may conclude that it is impossible to cross it. Helping students to bridge this gap and to retain interest and enthusiasm for using what they know, rests primarily with the clinical teacher and, unless she can find some way to do this, she cannot help her students to achieve independence and maturity in dealing with clinical problems.

Recognizing that the student is particularly vulnerable to these discrepancies and considering some of the reasons why this is so, will help the teacher to succeed in this task. Among the reasons for the student's vulnerability are the idealized view of nursing which many students hold, the difference between her own performance and that of her role model, and the discrepancy between the amount of influence she has in nursing situations and the amount she would like to have.

The student who enters nursing school usually views nursing as a way of righting injustice and easing suffering. No wonder, then, that she is dismayed and disillusioned when she finds that the nurse does not have as much power to right wrongs and to take away pain as she had thought. She soon discovers that even the most earnest application of available knowledge and the most dedicated nurses cannot ease everyone's misery. She also discovers there are injustices within the profession itself. Not all nurses put the welfare of patients first; resources are not always used in the patient's best interest, and so on. In addition, students often tend to see issues as either black or white, with little tolerance for various shades of grey; they are therefore likely to be distressed by the ambiguities which they observe. One of the teacher's tasks involves helping students to see the workable possibilities for change when they say, in effect, "This whole situation is totally wrong and it must be done away with right now."

Not only does the student encounter discrepancies between her ideal view of what nurses should be and what they are; she also is confronted with the fact that she herself does not invariably meet the high expectations she has set for herself in giving patient care. How humiliating, for example, to enter a patient's room ready to do a fine job of caring for his decubitus ulcer, only to discover that she must make a hasty exit five minutes later because the odor has made her nauseated! How disheartening to realize, after hearing so much about the nurse's role in supporting bereaved family members, that she swept past a patient's wife who was weeping in the corridor, because she could not bear to approach a person showing grief.

The student is also often disturbed by the difference between the amount of influence which she actually has over situations, and that which she expects to have. She may report to the head nurse and physician, for instance, that a patient seems despondent. She repeatedly reports and charts examples of the patient's behavior which indicate feelings of hopelessness and despair, but notes that no action is taken in relation to the patient's emotional state. One day the patient commits suicide, and the student, with understandable anguish blames herself saying, "But I should have found a way to *make* them listen." Similar, less dramatic incidents occur daily and can lead the student to feel quite powerless. When the chasm between what she herself can accomplish and what patients require is wide, it is difficult for her to accept the limitations of her influence and authority.

When students bring up examples of the gap between the quality of care which is given and that which should be given, it is important for the teacher to acknowledge that this discrepancy exists, and to consider some of the reasons for it as well as some ways the student can improve the care the patient receives. No matter how controlled the situation for clinical practice may be, the student often perceives that not all the factors which are operating are of benefit to the

patient. Some of our cultural values support improved patient care; others do not. Although there is a great deal of
discussion about the importance of having qualified nurses
give direct care to patients, society in many instances is
unwilling to pay adequate salaries to those who give this
care. Because health professionals form a part of this society
they are themselves influenced by society's positive and
negative values concerning health care, even though they
strive to counteract those forces which interfere with patient
care. For example, adults in our culture are expected to be
productive, self-supporting, and sufficiently in control of
their emotions to avoid burdening others. It is not surprising
then, that staff may gravitate toward a young, cheerful
woman who is recovering from surgery for the removal of a
benign breast cyst, and who is surrounded by gifts, flowers,
and cards from her family and friends. In such a situation
the "supports" are very evident: a favorable prognosis, concern of family and friends, and the patient's pleasing
personality which makes it enjoyable to spend time with
her. The staff may have quite a different reaction, though,
to an elderly patient who has had a radical mastectomy,
whose husband died recently, and who has no close family or
friends. She cries easily, expresses despondence and feelings
of helplessness. The acuteness of her loneliness, the dilemma
of how this woman, who has not worked outside her home
for many years, will manage to support herself, her expressions of grief, and the uncertainty about her ultimate prognosis can make it difficult to work with her.

As students bring up such dilemmas it is essential to
acknowledge the realities of the situations, and then to concentrate on measures which the student can take in caring
for the patient. Exhorting her to do what others obviously
find difficult to do, without making further suggestions
about how the student may go about caring for the patient,
is of little help. Unless students are helped to deal with such
situations they may resort to criticism of the ways others
relate to the patient, and to withdrawal of their own efforts.

Criticism of the patient's family, for instance, is a common way of diverting attention from what the student herself can do for the patient; sometimes she does not realize that the same factors which lead her to avoid the patient also may make it difficult for the family to maintain their help and interest.

Discussion of measures which the student can take is more useful and such measures are more likely to be applied when they are based on acknowledgment of the real difficulties which the situation presents. Instead of saying, "Spend time with the woman who had the radical mastectomy; she needs you more than the young woman who had a benign tumor removed," it would be more useful to help the student learn ways of dealing with the patient, and also how to protect herself, so that she does not become overwhelmed by the patient's despondence. Rather than saying, "Encourage the patient to cry" (which could result in the patient's feeling even more helpless, and in the student's feeling helpless too), it would be preferable to guide the student in helping the patient to express some of her concerns. Although the patient may cry as she speaks of her losses, the emphasis will be on the patient's concerns, rather than on the crying itself. The familiar terms "time, distance, screening" used for protection of staff when working with patients receiving radiation therapy can be applied when working with this patient. Because her feeling of loneliness and emptiness is very demanding, extended contact with this woman may be too taxing for any one nurse. Is it possible then for several nurses each to spend some time with her every day? "Distance" and "screening" may be achieved by planning periods when, in addition to her "one-to-one" contact with the patient, the nurse interacts with her in a group. Such measures can help in a difficult situation like this, which may require not only nursing intervention, but psychotherapy, and job training and placement, along with careful follow-up to determine the presence of further malignancy.

Sometimes nursing intervention seems too commonplace to the student, who would like to find some dramatic way to restore the patient's health and happiness. Perhaps one of the most difficult types of discrepancy with which students must deal involves situations in which no amount of dedicated nursing and medical care can surely and completely restore the patient's health. However, students can be taught that the dependable use of measures which lie within their scope can gradually enable the patient to help herself more effectively. They can be taught that their efforts to do what they can may encourage others to do likewise; for instance, other patients in the hospital may stop by for a friendly talk with the depressed patient when they notice that the nurse does this.

Besides recognizing the student's vulnerability to discrepancies between ideal and actual patient care, and acknowledging these discrepanies and helping the student learn to care for the patient, the teacher can assist her to learn some other useful ways of maintaining her ability to function in situations where there is a wide gap between actual and ideal care. One of these measures involves learning to protect herself from the impact of the discrepancy by channeling her energies toward the solution of the most urgent problems in a situation. The student may find it difficult to understand the necessity for such protection. It may seem to her that every staff member who does not battle each problem to the hilt, every day, is callous and negligent. It can be pointed out to her that on a day when the nurse may be faced with the laundry's failure to deliver enough sheets, a physician's forgetting to order a patient's diet, the dietitian's saying that because of inadequate help no delayed breakfast trays will be served, as well as the admission of a critically ill patient, she must channel her energies so the most important problems are taken care of. The student needs help in discriminating between issues of greater and lesser importance, and in differentiating between those problems which lie within the scope of the nurse's authority to handle, and those which do not. Otherwise the newly graduated nurse may resemble a bright meteor, full of fire to bring about change, but whose sparkle is quickly extinguished by discouragement and a feeling of powerlessness.

Another way in which the clinical instructor can help the student cope with the discrepancy between the actual and ideal involves helping her realize that patients can get well despite adverse conditions. The teacher recognizes the resilience of human beings which enables them to recover even when their care is not ideal, and she has also developed the ability to differentiate between the degrees of seriousness of various lapses in patient care. She knows, for instance, that failure to test one urine specimen is not as serious as giving a diabetic a double dose of insulin. To the inexperienced student, however, each evidence of imperfection in patient care may present a seemingly insuperable barrier to the patient's recovery.

While the beginner may overestimate the seriousness of one lack in the patient's care, she may fail to note another which is more important. For instance, she may be so concerned because there is no clean spread for a postoperative patient's bed, and so busy looking for one that she does not notice that he has taken no fluids for the past 12 hours, and that no one (including herself) has offered him any.

Although the student's lack of experience handicaps her observation in some ways, it also gives her a perceptiveness and freshness of insight which those who have worked in a situation for a period of time often lack. In the process of helping the student to gain perspective and recognize the relative importance of various facets of the patient's care, it is important for the teacher not to discourage the student's fresh viewpoint (which she tends to lose all too quickly) but to nurture it. On her first day in the psychiatric unit, for example, the student may notice that an elderly woman goes up to each staff member (and even to the student, who is a stranger to the patient) crying and talking in an agitated way about the day the ambulance came to take her to the hospital and about her terror and humiliation on that occasion. Although the staff takes this behavior in stride as part of the patient's illness, the student may be overwhelmed by the thought of what it must be like to be so upset that one shows so little discrimination in selecting people to talk with about personal concerns. Rather than discouraging this observation when the student mentions it in conference by a brisk, "Well, that kind of behavior can be expected of a patient in her condition," it is preferable to allow the student to discuss her reaction to the situation. Such encouragement from the teacher can help forestall some of the rather stereotyped thinking so prevalent in nursing, and can help the student to realize that her observations are worth sharing with others.

There is no conflict between educational control of clinical experience and helping students achieve independence. At no time does the teacher relinquish her responsibility for the selection of learning experiences to implement educational objectives or for the provision of the direct supervision required by students. Her selection of experiences is as deliberate and careful for the senior as for the freshman. As the student progresses in the program, however, it is important that the teacher emphasize the selection of clinical experiences which will help the student increase her sense of responsibility and her independence.

Some Problems in Evaluating Clinical Performance

Evaluation of clinical performance is a subject about which much has been written. It would be presumptuous to suggest that there are any definitive approaches to this procedure, which all clinical teachers find difficult and baffling. It may be helpful, however, to raise some questions for consideration and to share some thoughts on this subject.

Objectivity and subjectivity

Teachers are often advised to be objective in evaluating students' performance—to keep themselves out of the evaluation process as much as possible. It would seem, however, that exclusion of subjectivity from clinical evaluation is impossible, because the process involves interaction between student and teacher. One undesirable result of the emphasis on objectivity is reluctance to consider the ways in which one's own interaction with the student affects the process of evaluation. Another is insufficient appreciation for the positive and useful effects of the teacher's subjective impressions of students' performance. These subjective impressions give the teacher "hunches" which, while she should not act upon them unless they are verified by examples of students' performance, can serve to alert the teacher to aspects of nursing care with which students may need assistance, as well as to types of care which students may carry out especially well. For example, the teacher may suspect that a student is afraid of a stroke patient who has sudden bursts of temper, and is therefore having difficulty working with him. Although the teacher does not yet have any concrete examples of performance with which to validate her "hunch," further observation of the way the student cares for the patient and further discussion with the student may serve to confirm her impressions and lead the teacher to find ways of helping the student work more effectively with the patient. Or, the teacher may have a "hunch" that a particular student is developing unusual skill in supporting mothers during labor, and this may be verified after further observation of the student's care of these patients. Who can explain the basis for such hunches? Possibly it is the tone of the student's voice, her facial expression, or something about the way the patient responds to her which registers with the teacher, giving clues which can prove useful in her future work with the student. Instead of trying to avoid and to discount subjective impressions, we should acknowledge them, use them as productively as possible, and seek ways to safeguard them from becoming harmful bias.

Safeguards against bias

The chief way to prevent subjective impressions from resulting in bias is to differentiate between one's hunches or speculations and concrete observations of performance which can be validated with the student and with other teachers.

Clarity about the performance expected of students is another important safeguard. For instance, what do we consider satisfactory performance after four laboratory periods devoted to practice in taking vital signs? Is each faculty member expecting a similar level of performance? Subjective impressions are less likely to lead to bias in evaluation when the teacher concentrates on assessing how well the student meets expectations in relation to each aspect of the course at various periods during the course. Some types of performance lend themselves more readily to specific description than others. It is easier to identify each correct step in taking and recording body temperature and cleaning the thermometer than it is to specify effective steps in leading a group discussion. One teacher may characterize the way the student led a conference as "over-directive" while her colleague views the same performance as "giving essential guidance and providing necessary information." They would not disagree, however, about whether the student's reading of a thermometer was correct. Although the teacher must try to identify the behaviors expected in all aspects of nursing practice, she must recognize that the description of expected clinical performance has a greater degree of concreteness and specificity in some areas of nursing than in others. In those areas in which descriptions of expected performance are less concrete, there is greater opportunity for bias and greater need to use other safeguards to counteract it.

Validating impressions with a colleague provides another safeguard against bias. In team teaching, one can compare impressions with other teachers involved in the same course; when one teaches alone, validation can be sought from those who taught the student previously. It is also useful to validate one's observations with the student. This can be done by reviewing with her, frequently and regularly, one's observation of her clinical performance. Such reviews provide necessary guidance for the student and also serve as a check on the accuracy of the teacher's observations and on the validity of the conclusions she draws from these observations. One teacher, for example, observed that a patient had not been assisted to walk, although the physician had left orders the previous day that this be done. When she mentioned this to the student, the student clarified that the doctor had just visited, had asked that the patient be kept in bed, but had not yet written the new order on the patient's chart. What might otherwise have appeared to be forgetfulness on the part of the student proved instead to be careful adherence to the physician's instructions. Such discussions with the student can also lead to discovery of areas with which she needs further help. For example, one teacher said, "I notice you did not roll up the patient's knee gatch," whereupon the student quickly said, "Oh, I would have except that the handle to raise the gatch doesn't work." What seemed to be a good application of information from the lecture on measures to prevent thrombophlebitis was actually failure to apply the information, which did not become evident because the gatch did not work.

Sharing information about criteria for evaluation

The criteria being used for evaluation must be made as clear to students as possible. Saying "Here is what we expect you to do when you give an intramuscular injection" and then describing the expected performance makes clear what criteria are used. If team teaching is employed it is essential that all members of the faculty and all students in the course meet together for discussion of criteria to be used in making evaluations; otherwise, it is possible that faculty members will disagree on some criteria without realizing it. Prior to meeting with students, of course, the faculty meet alone to discuss the criteria to be used, and come to some agreements. However, it is a common experience for faculty to spend many hours discussing criteria for evaluation and to think they have reached agreement in all major areas only to find later that additional areas remain to be explored and clarified. Also faculty members do not always agree on the relative importance of various criteria. For instance, two teachers may agree that students should teach patients, and also agree on the criteria for evaluating effective teaching, but one instructor may place greater emphasis on this aspect of patient care during clinical practice than does her colleague. Faculty also differ in their views about how a particular clinical situation (such as the timing of injections of Demerol for a particular postoperative patient) should be handled. No matter how many discussions of criteria for evaluation are held by the faculty, it seems that, in reality, students must be expected to deal with and learn from the different approaches and emphases shown by the faculty during clinical work. The important point here is that these differences be explored as fully as possible among faculty and that those which persist after discussion be made as explicit to students as possible. Such an approach helps students to profit from the varied viewpoints of faculty, and by observing the differences in the way faculty function, they become more aware of the variety of acceptable ways which exist for handling a clinical problem. This also demonstrates to students that differences in viewpoint can be openly discussed, and that those who disagree on some points can still respect each other and work together.

Sharing evaluation of student performance
with other teachers

Faculty members must decide the extent to which they will share their evaluations of students' performance with each other and also how they can do this without interfering with continuity of instruction or with each teacher's freedom to form her own impressions of students' performance. Is it a good idea, for example, to read the evaluations written by the previous teacher and confer with her about students' performance? Or is it preferable to begin working with students without this information? Some teachers maintain that they want no prior information concerning students' work, lest their own evaluation be colored by it. Others say they need this information, can use it prudently in working with students, and can avoid letting it influence them when they evaluate students' performance. Without some communication among faculty, each teacher develops hypotheses and ways of working with each student which she does not share with her colleagues. How many weeks then are required before another teacher recognizes a problem previously identified by her colleague, only to find that it is almost too late to help the student with it, since the course is almost over?

Some faculty members prefer a middle ground—that is, they like to start working with students without information concerning previous evaluations of the students' work. If the teacher encounters problems or difficulties, she then speaks with a colleague who taught the student previously, sharing and validating her impressions and noting areas of disagreement. Still another approach is to have a conference with the student's previous teacher at mid-semester in order to discuss impressions of each student's performance. This allows one to validate one's initial impressions well before the end of the course. The last two alternatives have particular merit because they permit the teacher to make an initial evaluation of the student's work without the possibility of being influenced by another teacher's views, and to validate her impressions with a colleague.

Disagreement about evaluation of students' work may, if it concerns the entire group of students, reflect the use of varying criteria for evaluation. If it concerns one student in a group, it may indicate that a particular faculty member is having difficulty in working with a particular student. In the latter case, noting how a colleague responds to the student's work is one way of coming to a better understanding of the basic reason for the difficulty the teacher has in working with a particular student.

When more than one teacher has provided clinical instruction for the same group of students during a course, the faculty often share their impressions of students' performance not only in order to provide continuity of instruction, but also as a means of collaborating on preparation of written evaluations. This can be a satisfying process which further clarifies each teacher's impressions of students' performance and helps counteract bias. However, it can also lead to submerging the differences in each teacher's impressions for the sake of preparing a written statement on which both faculty members agree. When there is real discrepancy in the way teachers evaluate a student's work and if this discrepancy persists after discussion, it is preferable to indicate this fact on the written evaluation, rather than to formulate a statement which is so general that it does not express either teacher's view. To admit differences of opinion when writing evaluations reflects a belief that neither opinion is necessarily completely wrong or completely right. They are simply each teacher's impressions, which she supports with examples of the student's behavior, and with which her colleague, who has observed the student under different circumstances, may or may not agree. Stating in written evaluations the differences in the views of various teachers can provide the basis for useful observations over a period of time concerning the way the student performs. When reviewing a student's records over two years, for instance, it may appear that she consistently performs better when working with relatively inexperienced teachers than with those who are chairmen of departments. Such an observation can lead to the possibility of discussing this observation with the student in order to help her to consider possible reasons why she seems to work more effectively with one faculty group than another, and what steps she may take to cope with the situation in the future.

Considering the varied complexity of clinical situations

It is difficult to provide situations of similar complexity to use as a basis for evaluation of clinical performance. When evaluating students' skill in taking blood pressures, for instance, it is important to remember that some patients' blood pressures are easier to obtain than others, and that some sphygmomanometers are easier to use and more accurate than others. The varying complexity of some other clinical situations is even more difficult to delineate. Is it easier for any student to teach Mrs. Jones than Mrs. Brown, or is it more difficult for a particular student to teach Mrs. Brown, because of the way the student and patient interact with each other? Raising such questions does not, of course, solve the problem. It does suggest, though, that our discussion of the evaluation process take account of these variables, and of our inability to completely control them; otherwise we can convince ourselves that evaluation of clinical performance is based on more tightly controlled situations than is actually the case. Realization of the difficulty in controlling the complexity of clinical assignments when students' performance is being evaluated need not lead to slackening of effort to deal with evaluation in a disciplined and knowledgeable way. In fact, it can enhance the process, by helping teachers to avoid ascribing greater reliability to the evaluation process than is warranted.

Separating evaluation from teaching and practice

Nurse educators have stressed the importance of not confusing teaching, practice, and evaluation. The teacher demonstrates for example, the taking of vital signs, provides practice periods with supervision and, finally, evaluates each student's performance. This method has done a great deal to clarify the differences between the processes of teaching, practice, and evaluation, and has emphasized the necessity for providing periods of practice before carrying out evaluation for grading purposes. It can also create problems for the teacher when she tries to separate teaching, practice, and evaluation from each other.

One of these problems involves the extent to which faculty is able to separate the process of evaluation from clinical supervision. Teachers differ in their ability to do this, but probably no instructor is impervious to "registering" impressions of student performance during practice periods, however often she reminds herself that evaluation properly comes later. The fact that Ann Doe consistently tangles herself in the sheets when she attempts to fold them while Rose White manages to fold them deftly each time, comes to the attention of the teacher in a way which may be described as "unofficial" evaluation. In order to eliminate this problem it may be well to have the evaluation done by another teacher—one who did not supervise the students' practice. Although regularly scheduled separate periods for evaluation are useful in making clear just when evaluation is taking place, this method also makes it difficult to evaluate care which is given over a period of time, with each experience building on the preceding one. When a student works with a geriatric patient over a period of several months in order to help him become independent, at what point does one say practice stops and evaluation begins? In such instances, it is necessary to evaluate the process as it unfolds, as well as the final result, if one is to help the student progress satisfactorily from one step to another. Perhaps the question really is how to separate the evaluation which is a necessary part of teaching and supervision from that which is done to arrive at a grade. The student who is practicing with a patient is not "just practicing" as would be the case in a situation where no patient is involved. Whenever the student works with patients, evaluation by the teacher of the safety and effectiveness of her practice must be carried out at the time in order to help the student improve her performance, as well as to safeguard the patient. Whether or not separate periods are scheduled for clinical evaluation, it is important to help students recognize that their performance after graduation will most likely be evaluated concurrently with their clinical work, rather than at separate periods designated for evalution. Evaluation of her work as a practitioner will probably be based on the principle that "everything registers; everything counts." A particularly good piece of work, a poor one, or a mediocre one is likely to be noted by her supervisors as her work proceeds. It is important, therefore, to help the student avoid developing an attitude of "This time doesn't count; this is practice," but instead to recognize that every aspect of care counts, as far as the patient's welfare and her own progress are concerned, and that it is being evaluated in order to protect the patient as well as to help her improve her performance, even though the grade she receives may be based on a separate evaluation.

Such an approach has the advantage of realism, and is therefore effective in preparing the student for her future work and for other life experiences too. How easy it is to wish that one could somehow erase certain actions, so as to escape their consequences. Nevertheless, it is as true for our students—as for the rest of us—that every action carries consequences, and that there really is no such thing as temporarily turning off consequences, even when one is a learner. The way a student deals with a patient one day sets the stage for the care she gives subsequently and helps determine the effectiveness of this care. While emphasizing this continuously operating interplay of cause and effect, it is essential also to convey to the student that, because she did poorly one day, she has not been irretrievably pegged as a poor student. The effective teacher conveys to students that even though they perform inadequately on one occasion, they have opportunity to do better the next time, and that she firmly expects them to do so. To the student who says, "But I could do better if only I were not being evaluated—evaluation makes me nervous," the teacher can reply, "But life is not like that. We are all being evaluated all the time and this evaluation affects our success at work and in other aspects of our lives. The results we ultimately achieve are determined by what we do each day."

Encouraging self-evaluation

There is general agreement that students should be encouraged to evaluate their own performance. It is therefore important to consider measures which foster self-evaluation. Procedures for self-evaluation are least effective when they are handled in a rather routine fashion, such as, for example, handing a student a rating scale and asking her to check her own evaluation in red ink, after which the instructor will make check marks in blue ink. Sometimes the student is asked to fill in the rating scale first—a procedure which sometimes results in the student's spending considerable time speculating on just where the instructor is likely to place check marks, and the relative merits of placing her own checks somewhat lower (for modesty's sake) or as nearly in the same location as possible. Such an approach to self-evaluation really becomes a guessing game, and a rather unfair one if the student is asked to "show her hand" first. I have deliberately exaggerated the negative aspects somewhat in order to illustrate some of the problems encountered.

Whatever the procedure of recording the student's self-evaluation, the mechanics are secondary to two other considerations. The ability and willingness of the student to evaluate her own performance depend largely upon her relationship with the teacher, and on her clarity concerning criteria upon which evaluation is based. Sometimes self-evaluation is spoken of as though it were easy. Upon reflection, one realizes that self-evaluation is a very difficult task which most adults have only partially achieved. Therefore it seems unrealistic to expect undergraduate students to be highly effective in evaluating their own performance, particularly when the person with whom they are expected to carry out the evaluation is in a position of authority and is responsible for assigning a grade.

Despite these limitations, however, students can be encouraged to evaluate their own performance. Individual conferences can do a great deal to assist students gradually to feel secure enough to talk with the teacher about what they believe their strengths and weaknesses in clinical performance are, without a great deal of apprehension that the teacher will allow these discussions to influence negatively her own evaluation of students' work. Sharing by teacher and student of the way each assesses the student's performance helps both of them to note areas of agreement and disagreement, and provides opportunity to discuss further the points upon which they do not agree. These discussions often serve to elicit additional information about the student's work which increases the validity of the evaluation process.

The teacher who shows students that she evaluates her own effectiveness, and that she seeks their assistance with the process, provides an example for students in carrying out self-evaluation. For instance, asking students to evaluate the effectiveness of classes, conferences, and various clinical experiences, and being appproachable when students wish to discuss problematic situations in relation to one's teaching, help students to understand more fully the process of self-evaluation.

Some conflicts between idealism and actuality

Some of the most disturbing aspects of evaluation involve the conflict between what most of us wish were so and what actually is so concerning the performance of students we must evaluate. We know that it is not always the student who tries the hardest who achieves the highest grade, but where is the teacher who has not wanted to give that "A for effort" to a student who has tried especially hard, even though her work barely merits a B!

This problem is one we share with teachers in all fields. However, it appears to be harder for clinical faculty and their students to solve because of the different frames of reference within which one works with students and with patients. In other words, evaluation of students' progress in an educational setting, and of patients' progress in a clinical setting, proceed on very different premises. As students and faculty move back and forth between patient-care and educational settings, it is necessary for them to recognize the differences in orientation to evaluation which are appropriate for each setting. One does not set hard and fast goals for patients, to be achieved within a specified period of time. For instance, there is no definite point in his progress where the laryngectomized patient who is learning esophageal speech passes or fails, in the eyes of those who are caring for him. He is simply expected to do the best that he can within a flexible time schedule. When it comes to evaluating students' performance, however, time limits *are* imposed, and minimum standards for acceptable practice have been set. Sometimes these distinctions are not made clearly and openly enough, and attitudes appropriate to the clinical situation are unwittingly carried over, by students and occasionally by faculty, into the educational sphere, thus leading to misunderstanding concerning the premises on which students' evaluations are based.

Teachers differ in the relative weight which they place on the progress the student has made in developing skills, the amount of effort she has expended, and the extent to which she has fulfilled the criteria for evaluating work in the course. One teacher or team of teachers may place almost exclusive emphasis on the extent to which criteria are met; the grade they give is not greatly influenced by the amount of progress the student has made since the beginning of the course, and the amount of effort she has put forth. Others are influenced considerably in their grading by the effort the student makes, and the amount of progress she has made in developing her skills. Although the extent to which these factors should influence students' grades is debatable, clarity in explaining to students the basis on which they will be graded is essential. If students are being graded almost solely on the extent to which they meet stated criteria, it is not fair to imply to any student that the amount of effort she expends will substantially affect her grade. It is easy to fall into the trap of doing this without realizing it, however, in an effort to encourage a slow student or to motivate one who does not seem to be exerting sufficient effort.

We shall never, of course, prevent all misunderstanding about grades. There will probably always be students who say, "But I worked so hard that I deserve a better grade." However, by honestly stating the extent to which various factors influence grades we can forestall some of the confusion, disappointment, and occasional bitterness which students experience in relation to evaluation.

Evaluating achievement in the total curriculum

So far, we have considered evaluation in relation to a student's performance in a particular clinical course. In addition, it is necessary to evaluate her achievement in the total program and to discuss with her the areas of particular strength and weakness in her performance in relation to her professional interests and goals.

Few students perform equally well in every clinical area. One of the concerns of faculty in relation to evaluation is the extent to which students should be expected to be "well rounded" in their clinical interests and skills. In the past, a great deal of emphasis has been placed upon the importance of versatility in nursing. Students were encouraged to think of the good nurse as being able to step into almost any nursing situation and handle it adequately, even though her above-average performance might be limited to one field of nursing. Of course, part of this attitude had to do with the comparative lack of nursing and medical knowledge in years past. When nursing consisted largely of assistance with personal hygiene and nutrition, and medical therapies were relatively uncomplicated, it was more reasonable to expect a nurse to be able to function adequately with many different categories of patients. As medical therapies have become more complicated and nursing care more complex, the need for specialization in nursing has steadily increased. As an example, a nurse who has not had special preparation is not expected to care for the patient who has had open heart surgery.

As specialization becomes more and more necessary the fields of specialization become narrower. The medical-surgical nurse who works with convalescent patients is not expected to be skilled in working with patients in an intensive care unit, for example, because this type of nursing has become a specialty in itself. Our students are surrounded by specialization and many of them are thinking concretely about their own future field of specialization.

One desirable consequence of the trend toward specialization is that it has encouraged nurse educators and students to be realistic in their thinking about individual differences in interest and ability. It has become acceptable nowadays (and rightly so) not only to say, for instance, "I'm interested in specializing in care of premature infants" but also to add, "And I really am not good at geriatric nursing, and I don't like it." Such a comment from a student 25 years ago would probably have sparked a lecture on how important it is for the nurse to enjoy working with all types of patients. Of course, each student must perform acceptably in every clinical area, but today she is quite free to acknowledge her preference for one field rather than another.

Have we, though, in the process of accepting individual differences and encouraging specialization, neglected to help undergraduate students develop greater competence in fields which are not their major areas of interest? It appears that we have, in some instances. Although we should respect each undergraduate student's gifts and interests, and not push her to excel in all kinds of clinical work, it is important to encourage a breadth of vision and a concern for various fields of nursing. Sometimes a student's interests are exceptionally narrow and, provided she passes all her courses, little attempt is made to help her broaden her professional outlook. Such narrowness may be an indication of an emotional difficulty which it would be worthwhile for the student to investigate, so that her career does not become unnecessarily restricted. Can the student who is interested only in pediatric nursing, and who does not like working with adults, be helped through counseling, to learn some of the reasons for this reaction? Can she, perhaps, be helped to learn to work more comfortably with adults during her student days and, although specializing in the care of children, maintain a professional interest in the care of adults after graduation? If so, her career will be enriched. Not only will she have sufficient interest to talk knowingly with nurses outside her own field and to read more widely, but her practice with children can be strengthened by noting how aspects of other fields of nursing have relevance for her specialty.

Breadth of interest must be fostered during the undergraduate years, because for most nurses the emphasis after graduation is on selecting one area of nursing and then on concentrating their efforts on refining their knowledge and skill in this field. Teachers whose interests are broad are understandably in the best position to help students to enlarge their scope and to see the relevance of various principles in a variety of fields and settings. However, each faculty member must necessarily be concerned primarily with the student's performance in the particular course she teaches. In order to place the evaluation of the student's achievement in the larger context of the total program, someone must evaluate (with the student) her total performance in the program, and consider her areas of particular strength and weakness as well as her special professional interests. This role may be assumed by a faculty committee,

158 THE PROCESS OF CLINICAL TEACHING

the student's faculty advisor or an individual faculty member.

Conclusion

Clinical teaching involves a constant interweaving of theory with its application to patient care, and the provision of meaningful opportunities for students to acquire skills necessary to make this application. It is a process which must be carried out with consideration for the needs of patients, students, and agency staff in each nursing situation. Because clinical experiences are rooted in the reality of the patient's nursing requirements, they form the heart of the study of nursing. Using them well is a challenge to the clinical and instructional judgment and skill of the nurse teacher.

SUGGESTED READING

Clissold, Grace K. *How to Function Effectively as a Teacher in the Clinical Area.* New York, Springer, 1962.
Conant, Lucy H. "Closing the Practice-Theory Gap," *Nursing Outlook, 15*:37, November, 1967.
Heidgerken, Loretta E. *Teaching and Learning in Schools of Nursing.* 3rd ed. Philadelphia, J. B. Lippincott, 1965.
Ingles, Thelma. "On Developing Skilled Practitioners," *American Journal of Nursing, 60*:1482, October, 1960.
Jimm, Louise R., and Fine, Jerry. "A Shared Experience in Leadership," *Nursing Outlook, 15*:36, October, 1967.
Palmer, Mary Ellen. "Self-Evaluation of Clinical Performance," *Nursing Outlook, 15*:63, November, 1967.
Quint, Jeanne C. "Hidden Hazards for Nurse Teachers," *Nursing Outlook, 15*:34, April, 1967.
Rines, Alice R. *Evaluating Student Progress in Learning the Practice of Nursing.* New York, Bureau of Publications, Teachers College, Columbia University, 1965.

Part III

SPECIAL ADAPTATIONS OF CLINICAL TEACHING

7

Teaching the Atypical Student

The image a profession has of itself involves not only the kind and quality of work performed, but also the characteristics of those who enter the profession. During recent years the group of people entering nursing has become much more diverse than it formerly was. Factors that have brought about this change include:

• A general movement toward lack of discrimination in selecting applicants for education and employment.

• The growing recognition that exclusion of nursing-school applicants for such reasons as age, sex or race deprives them of opportunity and deprives society of the contribution they could make to nursing.

• The persistent demand for greater quantity and quality of nursing services.

• The realization that limiting the recruitment of nurses to young single girls intensifies the problems associated with interrupted practice and part-time practice, since most nurses recruited from this group will later combine responsibilities to home and family with professional activities.

• The increasing tendency to place nursing programs in colleges and universities, which tends to lead to the admission of a more diverse group than usually enters the diploma schools (for example, many associate degree programs have a particularly diverse group of students).

• The increase in geographic mobility, with the result that students from other parts of this country and from other lands are likely to be found in schools whose enrollments were previously limited to residents of nearby communities.

The recognition that persons with certain types of disability can make a contribution to nursing, and the consequent admission of applicants who formerly were likely to be denied admission to nursing school for health reasons.

The teacher of nursing is, therefore, likely to find that her class contains a sprinkling of "atypical" students. The factors which cause these students to be atypical differ with each school, depending somewhat on the type of program offered and the location of the school. The teacher may wonder how to go about teaching a student who is twice her age, or whether she will be embarrassed when demonstrating certain procedures in the presence of a male student, or how students of different races or nationalities will respond to the experience of studying together. Regardless of the groups represented by her students, the teacher whose classes are heterogeneous will find it useful to give consideration to the reasons for the various types of behavior exhibited by her students.

THE STUDENT WITH AN ATYPICAL EDUCATIONAL OR CULTURAL BACKGROUND

The teacher of nurses can no longer assume that all of her students will have attended an American high school at a particular point in time. She will need to take into account the differences in her students' educational and cultural backgrounds. The foreign student may be accustomed to a more formal student-teacher relationship, and to less emphasis on extracurricular activities than is usual in American schools. The values emphasized in different cultures also affect the way the student performs in the nursing program. One student was in frequent difficulty clinically because of her apparent lack of concern for patients' welfare. After working with the student for a time, the teacher realized that cultural differences were playing a large part in causing the problem. The student came from a country where people go about their activities in an unhurried fashion; bustling about is considered rude and undignified. When the student responded in a too-leisurely manner to the teacher's request to prepare a patient for the operating room at the scheduled time, difficulties arose which affected the patient's welfare as well as the student-teacher relationship. The student and teacher discussed some of the cultural differences between her country and this one, and the effect these differences might be having on the student's clinical work. They talked about situations in nursing which require speed and promptness, and about those in which an unhurried approach is an asset. Although the student had difficulty functioning in situations requiring speed, she was outstanding in her ability to converse gently and in an unhurried manner with patients and their visitors. She gradually developed greater awareness of how her cultural background was affecting her nursing practice in this country, and increased her ability to differentiate situations which required speed from those which could satisfactorily be handled in a leisurely manner. The student continued to function best, however, in situations which did not place a premium upon working rapidly.

The teacher may question her ability to instruct students who do not fit the usual description of student nurses. Every student-teacher relationship is based upon the teacher's greater knowledge of the subject and her ability to share this with the student. Because a student has five children does not mean that she knows how to give intramuscular injections, or that she understands the rationale behind a low sodium diet; she requires instruction which is based upon the difference between the teacher's expertness in nursing and her own. This difference does not necessarily extend to other areas of living.

Help her to overcome feelings of insecurity

Each sub-group of atypical students may feel insecure,

but for different reasons. The older student may wonder whether she will be able to adapt to her studies quickly enough to keep up with her younger classmates; the Chinese student may worry whether her proficiency in English will be sufficient to enable her to pass her courses; and the student from a restricted suburban environment may be ill-at-ease in her first contacts with students from other lands or of other races.

Students who may be thought of as "typical" may also experience insecurity when associating with classmates whose backgrounds differ from their own, because they are uncertain about what to expect from such "different" students. Readiness to become acquainted with people from diverse backgrounds differs with each individual and depends, to a large extent, on the previous opportunities one has had to associate with such others as well as on how strongly the atypical person in the group exemplifies his "differentness." Students may exclude a girl with a private-school background from their confidences for fear she may be "snooty." On the other hand, a girl from a wealthy suburban family will be able to associate more easily with her classmates if she has had previous contacts with others of different social and economic status, and if she is not so fettered by her upbringing that she is uncomfortable with her classmates and thus makes them uneasy in her presence.

In addition to concerns about success with the program and acceptance by teachers and classmates, students may wonder how agency personnel and patients will respond to them and how they will react to these persons. How will the white patient react to being cared for by a Negro student and vice versa? Will the head nurse on the pediatric ward welcome a male student?

When broad admission policies result in a heterogeneous student body, the educational administrators and faculty have a responsibility for dealing constructively with this diversified group. This may entail such specific activities as helping a foreign student arrange for tutoring in order to minimize her language handicap, thus showing conviction that a language barrier is not evidence of inferior intelligence, that it can be dealt with and that it is not something for which the student should be penalized.

Despite faculty efforts to establish uniform standards for evaluating all students' performance, the atypical student may find that she must surpass her classmates' performance in order to achieve comparable recognition for her work. She may be called upon to supply repeated proof that she is "nursing material." The world of work necessarily emphasizes the quality of what one produces. While it is generally accepted that accomplishment is the yardstick for measuring an individual's contribution to a work situation, the atypical student may find that the emphasis upon the quality of her work is so great that it excludes her acceptance as a person. Others may communicate to her, "You really are not acceptable, but your work is." A vicious circle may be established in which the student strives harder and harder to do an outstanding piece of work; these efforts are met with praise by faculty and classmates. But recognition of the quality of her work does not necessarily make up to the student for not being invited to join her classmates for lunch, or for not sharing confidences with a fellow-student. One of the most subtle forms of personal rejection involves over-emphasis upon the product, to the exclusion of the producer. Discerning teachers and fellow-students can, by being aware of this possibility, try to guard against it. They can convey respect and concern for the student, and a welcome to her as a person, in addition to recognizing the quality of her work.

The relationship between majority and minority groups of students can be very constructive, even though it may begin with marked feelings of insecurity on both sides. Meaningful relationships may develop because of differences between the members of the groups, rather than in spite of them. A white student and a Negro student may build a friendship which not only sustains them in their personal experiences, but which sheds new light for both of them on the relationship between the two races. Similarly, an older woman in a class is sometimes in a unique position to provide support and encouragement to her younger classmates, who may talk to her about some of their concerns which they might find difficult to discuss with a faculty member.

Nevertheless, the atypical student may have trouble in establishing such relationships. She may find that having come from a different neighborhood, or that having different manners or a strange accent are barriers to companionship with classmates. The majority-group students sometimes cite an external and seemingly insignificant aspect of the atypical student's grooming or manners as justification for avoiding her. Perhaps it is easier to concentrate on a tangible aspect of "differentness" than to acknowledge timidity about becoming acquainted with a person with a different background. Students may show such reactions more strongly toward a classmate than they would toward a patient, because a peer relationship carries with it the possibility of the intimacy of friendship. When students make a comment about a classmate from another land, such as "She isn't as clean as we are," it is important for the teacher to inquire what led them to this conclusion. In one such instance the student replied, "She doesn't shave under her arms." The teacher explained that shaving the axillae is not "de rigueur" in all countries, and that failure to do so reflects different customs concerning personal grooming, rather than evidence of uncleanliness. Raising such questions can help students to become more aware of cultural differences, and aid them to avail themselves of the opportunity to become acquainted with students of different backgrounds. The lowering of some of the barriers to relationships between minority- and majority-group students can also result in more occasions for the atypical student to associate informally with her classmates, thus gaining increased opportunity to observe the differences between their customs and her own, and then to choose those which she wishes to follow.

Recognize her individuality

The teacher's role involves cultivating an awareness of her responses to having a "different" student in the group, as well as perceptiveness about what this experience may mean to the student and to the other members of the class. In addition to showing acceptance herself, the teacher must provide the type of leadership which fosters acceptance of the atypical student by others. She can, for instance, share with hospital staff and other students her conviction of the appropriateness of the student's presence. When making plans for the student's experience, her manner should be definite and in no way apologetic or doubtful. She should not deny the existence of differences (such as the fact that a student is forty-five years old and a grandmother), but welcome them because of the enriched opportunities for learning which they provide for all students. It is important for her to recognize and respect the individual differences of each student in the group.

If the teacher has reservations about the appropriateness of the student's presence in the class, she is likely to indicate these reservations in ways which may escape her attention, but which are obvious to the student. Perhaps the teacher is "too nice" to the minority-group students, never criticizing their work, as she sometimes does that of the other students. She may become angry when an older student is absent one day due to the illness of one of her children, although she is not angry when other students miss two or three days because they are sick. All students should be expected to meet certain requirements, such as those concerning punctuality and regularity of attendance. The teacher must also have a certain flexibility, however, in dealing with lapses and mishaps which interfere with a student's meeting these requirements. Most people occasionally oversleep, or have car trouble, and so on. The teacher may respond in differ-

168 SPECIAL ADAPTATIONS OF CLINICAL TEACHING

ent ways to tardiness from such causes, depending on her attitude toward the particular student involved. She may think, "Mrs. Ross is late and it's all right because after all she has to get her son to nursery school." On the other hand, she may make sarcastic comments to a student, questioning the wisdom of attending school when she is responsible for the care of a child. Actually, what the teacher needs to do is to consider, with the student, the reason for lateness, the frequency with which it occurs, and what can be done about it; and this must be done in a way which does not convey to one student, or a group of students, that they are being held to different standards about punctuality than the rest of the class.

The teacher of a heterogeneous group of students also often needs to assist certain members of the group to develop plans of study. One student who had small children found that she could do her reading while the children were around, but that her written work, which required greater concentration, had to be done after the children had gone to bed. The student who comes from a home where there is little privacy for study may find that she must do her homework in the library.

Individual conferences help to clarify the working relationship between teacher and student. For example, a student's comment may make the teacher aware that she is using different standards in evaluating the work of certain students, or the student may indicate that justified criticism is being interpreted as evidence of prejudice. If the teacher can furnish examples of ways in which all students are held to certain expectations, the student may be helped to realize that the matter being discussed concerns the quality of her work, rather than her membership in a particular group.

Over-emphasis on such a particular difference as age or race, and lack of recognition of each student's individuality can lead to categorizing students according to a particular characteristic, and subjecting them to sets of expectations which may be wholly inappropriate. The older student who is expected to be slow may finish far ahead of her classmates because she has learned how to organize her activities effectively. The male student who is expected to be awkward in working with young children may excel in this because of his experience as a father.

Recognition of individuality includes anticipation and planning for the particular requirements of the atypical student. For example, the wise teacher will foresee that a separate locker room must be provided for the male students in a class, rather than waiting until an embarrassing flurry occurs at the agency when it is discovered that one locker room is no longer adequate. In addition to such highly practical matters as arranging for locker space, the teacher must be alert to the ways in which the student's "differentness" may affect the teaching-learning process. She may notice, for example, that an older student performs poorly on a multiple-choice test, although she participates intelligently in classroom discussion. Lack of recent practice in taking tests can place the older student at a disadvantage, but additional practice in test-taking can help her gain confidence and enable her to improve her scores. Differences in the way students of various ages learn must also be considered. The younger student can be expected to be quicker in such activities as taking timed tests. The older student can be expected to relate her greater experience in living to the understanding of nursing problems, and to be more self-directing and certain of her career goals.

Dividing students into groups for clinical assignments and conferences also must be handled with a view to the possible effect of such division on the atypical students. Two or three older students, for example, may enjoy being assigned to the same ward and conference groups. But such grouping should not be carried to the point where these students have little opportunity to work with others in the class. Because two students are Japanese, or older than their classmates, does not necessarily mean that they have other things in common, or that they enjoy each other's company.

Make use of her special knowledge or skills

The presence of atypical students in a class increases the likelihood that some students will possess greater knowledge or skill in certain areas that pertain to nursing, than the teacher has. A male student with knowledge and experience in electronics may be able to assist the teacher, as well as his classmates, toward greater understanding of electronic equipment used in the care of cardiac patients. Older women in the group may have had experiences that are especially useful in the study of nursing. For example, the experience gained in running a household can help a student to organize tasks involved in care of sick people. The student who has cared for babies of her own is not likely to hesitate to pick up a baby for fear of dropping him. Having experienced such events as the birth of children and death of loved ones can heighten one's understanding of what these experiences can mean to patients.

The atypical student should be encouraged to share her viewpoint and experiences with classmates during clinical conferences, thus contributing to a fuller discussion of many topics which concern nurses. Care of aged family members and characteristics of patients of different nationalities are examples of topics in which the viewpoint of men and women of various ages and of students from other lands can broaden the discussion.

There are ways in which the teacher can help the atypical student to feel at ease in the group. For instance, although discussion can be enriched by the contributions of a student whose background varies from that of the majority of the class, it is important not to expect such a student to take part in the discussion to the extent that she is made to feel conspicuous. It is helpful for both student and teacher to discuss their expectations about conference participation. The student may express a feeling of being caught between two fires: that of appearing a "know-it-all" in the group, and that of fulfilling what she thinks are the teacher's expectations because of her greater familiarity with certain topics. The teacher can help by indicating her recognition of the differences in the student's background from that of most of the class, and suggesting that she contribute in conferences to the extent that she feels at ease in doing so. Individual conferences can provide further opportunity for student and teacher to discuss aspects of the student's previous experience which relate to her study of nursing.

Diversification of the student group presents not only a challenge, but an opportunity for personal growth of both faculty members and students. Interaction with people of different backgrounds and characteristics is a stimulus not only to learning more about them, but also to learning more about oneself (by noting ways in which one is the same, or different, and considering possible reasons for these differences).

THE STUDENT WITH A CHRONIC HEALTH PROBLEM

Another type of "differentness" is presented by the student with a chronic health problem. The increased demand for nurses, and the growing variety of positions available in nursing (some of which can be handled by persons with disabilities) have led many schools to adopt more flexible policies than they formerly had concerning health requirements for admission. This trend has been increased by the placement of nursing programs in colleges and universities whose policies are often more liberal in regard to the admission of handicapped persons. The admissions officer may present the applications of students with some disability to the nursing admissions committee just as he would present the applications of such candidates for entry to any other division of the educational institution. The applicant's suitability for nursing must then be considered by the nurse faculty in relation to the requirements of the program and the type and severity of the applicant's disability. (For example, the student with well-controlled diabetes probably would have no difficulty in completing the program.)

If the school provides clinical experience in off-campus agencies, it is often necessary for the college physician and the physician responsible for personnel health at the agency to share information about students' illnesses which may require attention during clinical laboratory periods, or adaptations in the way a student's clinical experience is planned. For example, the diabetic student who needs a snack at 10:00 A.M. cannot be assigned to scrub in the operating room from 8:00 A.M. until noon without having a coffee break. The extent to which the student should be held responsible for informing the teacher about a physical handicap is a much-debated question. Although in some instances it seems essential for personnel of the health services of the college and the clinical agency to communicate concerning a student's health problem, the student should be encouraged to assume this responsibility. The age and maturity of the student, the degree of likelihood of serious consequences to her and to others if she fails to share information concerning her handicap, and the policies of each institution concerning health records, influence how much responsibility the student should be expected to assume. Information given to faculty and health service personnel can be misused in a way which discourages the student from assuming responsibility in reporting her health problem, and often leads her to suspect that information about her health is not being held in confidence.

When it is necessary to share information concerning a health problem, the student should be advised of this ahead of time, so that she knows what facts about it should be shared, with whom, and why. In some instances, for example, institutional policy requires that health problems be reported. The personnel health service of a clinical agency may ask whether the student has a history of drug allergy, for example, so that emergency treatment could be made readily available should it be needed while the student is at the agency.

The teacher should be especially careful when sharing information about a student who has received care for an emotional disturbance. Health personnel are not immune to prejudices concerning emotional difficulties; in fact, their sophistication in these matters can add a subtlety to such prejudice which makes it especially difficult for the student to combat. Information about a student's emotional problem should not be shared without the student's knowledge, and the necessity for sharing it must be carefully evaluated. If a student believes that a counselor and an instructor are discussing her emotional problem, it can damage her relationship with both of them. It is usually best for the student to discuss with the clinical instructor the facts which must be considered in relation to her clinical work. For instance, if she is having difficulty caring for men patients and has sought the assistance of the counselor, it would be advisable for the student to ask the teacher for temporary assignments to women patients rather than having the counselor make the request.

Clarity concerning the teacher's role

The teacher needs to remember always that she is the student's teacher—not her nurse or physician. She must resist the temptation to handle students' minor illnesses herself rather than referring students to the health service. She must also resist giving health counseling which is not coordinated with the plan of care initiated by the physician. For example, arranging for a coffee break for a diabetic student lies within the province of the teacher; advising her about what and how much to eat does not.

Clarity concerning one's role is important also in relation to the teacher's evaluation of the student's clinical performance. Confusing the role of nurse with that of teacher can lead to failure to assume one's responsibility for evaluation, and even to expecting others who have not observed the student's work to decide whether she is performing satisfactorily. For example, if a student does not seem alert enough (whether due to prescribed medication, or whatever the cause) to function safely in the clinical setting, it is the teacher's responsibility to discuss her observations with the student. An attitude of "Well, her physician must think it is all right for her to continue clinical practice, so who am I to object?" reflects confusion of roles and abdication of one's responsibility. The student's physician cannot be expected to predict, with unfailing accuracy, how a student's work will be affected by her illness, nor is he in a position to evaluate her clinical work. One student may be temporarily unable to function clinically because of difficulties which seem no worse than those of another student who manages somehow to perform satisfactorily in spite of them.

The necessity for avoiding confusion between the role of nurse and that of teacher is underscored by the fact that what is best for a student at a particular time is not necessarily best for her patients or her classmates. Despite one's efforts to remain squarely within the role of teacher, it is often difficult to restrain one's "nurse-self" from compromising experiences for other students, or for patients, while attempting to "ride out" a period when one student is performing poorly. Flexibility is required, since many students experience slumps in the quality of their work, and not all kinds of nursing care problems involve the possibility of adverse effects upon patients. Nevertheless, when a conflict arises between what may be desirable for the student with a health problem, and what is required by patients and other students, the teacher must be clear concerning her primary role, i.e., the instruction of all students in the group, and responsibility and concern for the welfare of patients assigned to her students. Within this framework she assists the student with a health problem to maintain satisfactory clinical performance.

Sometimes others contribute to the process of role-blurring. For example, students may find it hard to understand why the teacher cannot care for them if they become ill. The physician may automatically turn to a clinical instructor with instructions concerning the health care of a student who is his patient. "After all, you are a nurse" may be the attitude expressed by administrators, especially if they are not nurses, in an effort to persuade nurse faculty to provide health care for students. Then it becomes necessary for the nurse-teacher to clarify her role, to channel students' and physicians' requests to appropriate resources, and to work with administrators of the school in order to obtain necessary health services (including facilities for counseling) for students. One administrator of a nursing program who had diligently pointed out to university officials that her faculty should not be expected to provide nursing service for students, nevertheless discovered that plans were under way to place the nursing department and the college infirmary adjacent to each other when a new building became available. Her query, "Why?" met with some embarrassed glances, and later to a change in the building plan. The nursing administrator recognized that her faculty already faced a challenge in defining their roles, just by being nurses and having close clinical association with physicians and students, and that also having a college infirmary on their doorstep would only result in further role-blurring.

The differentiation between what is appropriate to the role of teacher and the role of nurse may be less clear-cut when one works with students who have emotional disturbances, than is the case with students who are physically ill. The distinction between academic counseling and counseling for personal problems may be difficult for the teacher to make, particularly since many of the clinical situations which the student encounters touch upon her own concerns with such matters as family relationships, illness, and death. The teacher's emphasis should be upon the clinical performance of the student. Although this does not exclude consideration of the student's personal reactions to clinical events, the emphasis should, for example, be on helping a student to recognize that she is having difficulty working with children, rather than on personal experiences which the student has had which may be contributing to the difficulty. If the student seems unable to modify her ways of working with children after the teacher has made suggestions, she should be advised to seek counseling to help her understand and cope with the problem.

Not only is it sometimes difficult to differentiate clearly between academic and personal counseling but, in some schools, the lack of resources for personal counseling may make it difficult for the teacher to confine herself to the sphere of academic counseling. In contrast, resources for care of physical illness are usually more plentiful. If the teacher is not clear about her role she may attempt to fulfill expectations which are inappropriate and which sometimes exceed the limitations of her time and skill, instead of clarifying her role with administrators and working with them to obtain the needed resources.

Emphasis upon maintaining her appropriate role should not, however, be carried to the point where the teacher cannot respond humanly and humanely by taking over the role of nurse in those occasional instances when her clinical skill can sustain a student until other provisions can be made for her care. In other instances the change of role may not be so definite; a teacher of any subject might be able to deal with the situation, but the nurse-teacher can handle it more effectively because of her professional knowledge and skill. An example of the former type of situation occurred when a student suddenly developed a high fever and convulsions. It was obvious that she required constant nursing attention, and that the nursing staff could not provide this. Several faculty members volunteered to alternate in providing the care until the student's condition had improved. More usual, however, is the latter type of situation in which the teacher remains with a student who has a sudden abdominal pain, until the emergency room staff can take over, or stays with a severely anxious student until she is calmer, and then helps her consider what steps to take in order to deal with the problem.

In responding to such situations the teacher must assess as carefully as possible what is necessary and appropriate in light of the circumstances, recognize the limits of her skill as well as the contribution her skill enables her to render, and realize that she is functioning outside her usual role. Such situations must be differentiated from those in which the teacher attempts to substitute for other personnel over a period of time, either because of her own lack of clarity about her role, or because of pressures exerted by others, such as physicians and administrators.

After an experience in which the teacher has functioned outside her usual role, both she and the student may have some difficulty resuming their usual relationship. The student who has received physical care from a faculty member, or who has, in a moment of distress, blurted out some personal concerns triggered by the clinical situation, may find it difficult to face that teacher at the next clinical laboratory. She may wonder, for instance, whether something she said under stress will be brought up during an evaluation conference in a way which might imply that the teacher now has less respect for her, and that the way she handles herself from now on will be carefully scrutinized. In such a case it is important for the teacher to assume an approach which does not deny the incident, but which assists the student to resume her former relationship with the teacher. The teacher can facilitate the resumption of the usual working relationship by conveying, either verbally or by her unspoken attitude, that the incident is closed, that she recognizes that most people occasionally lose their aplomb, and that the event will not be allowed to intrude upon the work situation nor lower her respect for the student.

While such experiences may increase the teacher's susceptibility to involvement in problems (such as that described above), they also can be especially meaningful and challenging because they give her opportunities to transcend her usual role and thus to achieve a rare degree of personal as well as professional fulfillment. They can enable her to pierce through the impersonality of relationships which is characteristic of our culture, and to sharpen her awareness of the joy which comes with having the skill and the compassion to help another person. For the student such an experience may be evidence that in a crisis her needs as an individual were considered important enough for the teacher temporarily to venture beyond her usual role.

Referring the student for health care

All teachers are expected to refer students for health care when a health problem comes to their attention. However, the nurse-teacher has a heightened perception of possible health problems, and can therefore refer students more promptly. The constant slight tremor of a student's hands which is scarcely noticeable to a teacher or art or of English may lead the nurse-teacher to suggest that the student seek medical advice, as may her observation that a student appears consistently preoccupied during clinical conferences. Her background as a health professional can also enable her to recommend referral in a way which does not cause the student needless alarm. Sometimes a teacher hesitates to refer a student, lest the student become offended, or the matter be viewed as "making mountains out of molehills." In many instances, however, the student is already aware of the difficulty, and relieved that someone else is concerned enough to help her find a way to cope with it. The teacher whose attitude is one of genuine interest in the student's welfare, and who respects the student's privacy and her right to accept the referral or not, will ordinarily find that her relationship with the student is enhanced, rather than impaired, by the referral.

Our nursing heritage is not an unmixed blessing, though, when dealing with students who have health problems. Some teachers may be hampered by nursing's traditional emphasis upon the necessity for glowing health among those who would be nurses. Teachers who believe that people who have physical ailments or emotional problems really do not belong in nursing may communicate this attitude to the student and make it difficult or even impossible for her to accept a referral for treatment. It is especially important for the teacher to be aware of her attitude toward physical and emotional disabilities among students, and to avoid conveying to the student that a particular problem renders her unfit for a nursing career.

Helping the handicapped student plan realistically for future work

The nursing-school applicant who has a disability which may make her unable to practice in certain areas of nursing should be made aware of this probability before she is admitted. As she continues in the program, she should be assisted to develop realistic goals concerning the type of nursing which she can practice after graduation. The student with a slight limp due to polio may handle successfully sixteen to twenty hours of clinical practice weekly; after graduation, however, she may find that the physical demands of working in an intensive care unit for forty hours a week are beyond her, but that she can function well as a school nurse. Similar guidance is required for students who develop some chronic health problem during their course of study.

THE PREGNANT STUDENT

Because of the increased number of married women studying nursing, the number of pregnant students carrying out clinical laboratory practice has also increased. The teacher and student should discuss how clinical practice will be handled, in light of the recommendations made by the student's obstetrician. Certain types of assignments are obviously hazardous for the pregnant student; for example, caring for patients with German measles, or holding patients while x-rays are taken. Policies concerning whether a pregnant student may continue her clinical work have become more flexible in recent years. Sometimes the policies of the school and the clinical agency differ concerning how long the student may continue with her laboratory practice. When this happens, the clinical teacher must refer the matter to the administrative officers of both institutions so that a satisfactory solution can be worked out. Occasionally some agency staff members do not agree with the policy of permitting a pregnant student to continue clinical practice. One solicitous intern, for example, excitedly ushered a student out of a patient's room saying, "You shouldn't be in there." Explanation of the policies can help staff members to function within them.

The clinical instructor must overcome any temptation to give health counseling to the pregnant student or to the student who has recently had a baby. The instructor of obstetric nursing, for example, should recognize that it is not her function to advise the student how soon to resume clinical practice after delivery of her baby, but that this advice should come from the student's obstetrician.

THE GENIUS AND THE "ODDBALL"

Two other types of atypical students are the very bright student and the "oddball" who has not become professionalized as rapidly as her peers. Too often the extremely intelligent student is not sufficiently encouraged to extend her reading and study beyond that expected of the average student and, instead of developing her abilities, she may, to her own loss and that of the profession, study less than her classmates because she can achieve good grades without much effort. Rather than leveling off her academic and clinical achievement to correspond with that of her classmates, the student should be guided in wider reading and in undertaking such challenging clinical assignments as helping a withdrawn youngster who refuses his meals, to eat, or assisting with care of a patient following open heart surgery. It is important, though, to avoid consistently assigning particularly challenging or interesting clinical work to the brighter students, since this may deprive some others of sufficiently varied experiences, and lead to jealousy among the students. Word spreads quickly when an experience which students regard as a "plum" becomes available on the ward. For example, all students in a class may wish to care for a newly admitted patient on a Stryker frame. The teacher who has challenged her brighter students in an unobtrusive way will not have created an environment in which the student grapevine buzzes with "Louise will get to do it. She gets all the interesting assignments." Such a situation can be avoided by guiding the bright student in ways which do not attract other students' attention. The above-average student may be expected to accomplish more when working with a particular patient than would be expected of her classmates. She may, for example, work out a program of alternating rest and activity for an arthritic patient, although most other students in the group are expected, at that point in the program, to concentrate primarily on assistance with personal hygiene. Such modifications of assignment are not as obvious as repeatedly selecting one student to do the first gavage on an infant, or to make the first home visit to a patient recently discharged, or to be the first to observe renal dialysis or cardiac catheterization. Suggestions for additional reading can also help the able student to achieve considerably more than her classmates without being singled out. Why not encourage the able student to read widely in such works as those by Jung, Freud, and Sullivan during her course in psychiatric nursing, even though her classmates are not expected to read so extensively?

There is a fine line between encouraging a student and pressuring her. When a student shows aptitude for the teacher's own specialty, it sometimes requires restraint not to push her in an effort to help her get ready to practice in a field of nursing for which she seems particularly suited. (What teacher does not cherish the idea, whether she admits it or not, that her subject is really the most interesting and important—and that a student who shows particular appreciation and aptitude for it is a little bit special?) The important thing is that the bright student receive encouragement to achieve more than her classmates without having her course requirements raised so that she is pressured to achieve more.

The "oddball" student may, like a throbbing accompanying infection, cause discomfort which draws attention to a situation requiring change. This type of student asks questions which are sometimes too searching for comfort, and in so doing holds up a mirror in which her teachers and classmates may see an aspect of the profession which startles and even offends them. This is the student, for example, who, when asked to remove her class ring during clinical laboratory, inquires whether microbes also cling to wedding bands, which she notes are permissible. To be sure, some of the student's comments may be merely irritating, rather than enlightening. Nevertheless, by attempting to silence her questions and comments, her colleagues may lose opportunities to see issues in a different light, and to search out some of the contradictions which hobble the profession's efforts to improve nursing care. Judgment is required in fostering the student's original and useful insight. At the same time, limits need to be set on her outspokenness; otherwise she will antagonize others to the extent that they are unable to accept her contributions. This student should be helped to differentiate between comments which are merely irritating and those which, though possibly annoying to others, can stimulate discussion. She should also be guided concerning where and to whom she expresses her ideas. Discussing some of her unconventional ideas and criticisms with the instructor can be the means of channeling some of her comments; it can also help the student learn how to make suggestions to appropriate personnel in such a way that they are likely to receive consideration. Despite these efforts to assist the student to handle her comments constructively, it is not always comfortable for members of the profession to have some of their cherished ideas scrutinized. The challenge lies in tolerating the discomfort in order to profit from fresh insights.

Because this type of student is difficult to teach, she is occasionally labelled emotionally disturbed in order to justify avoidance of the challenge of teaching her. The student who is a "thorn in the flesh" is not necessarily emotionally disturbed; she may merely be more perceptive and more honest than her classmates. Of course, she may also have emotional difficulties which are reflected in relationships which seem always to result in irritation of others, rather than collaboration with them. It is important, though, not to come to this conclusion merely because the student is sometimes irritating, or asks incisive questions.

Conclusion

The fact that a student has lived longer than her classmates, that her skin is a different color, or that she has some physical handicap may stand out at first, but each student is different in her own particular way—such as being unusually shy or bright or creative. The teacher has the opportunity to help each student make a contribution to nursing by capitalizing on her uniqueness as an individual—a uniqueness which is not only recognized but nurtured while she studies nursing.

SUGGESTED READING

Anastasi, Anne, Ed. *Individual Differences.* New York, John Wiley and Sons, 1965.
Bibby, Cyril. *Race, Prejudice and Education.* New York, F. A. Praeger, 1960.
Cooper, Russell M., Ed. *The Two Ends of the Log.* Minneapolis, University of Minnesota Press, 1958.
Francis, Gloria M. "A Minority of One," *Nursing Outlook, 15*:36, June, 1967.
Giles, H. Henry. *The Integrated Classroom.* New York, Basic Books, 1959.

184   SPECIAL ADAPTATIONS OF CLINICAL TEACHING

Hughes, Everett C. *Men and their Work.* Glencoe, Ill., The Free Press, 1958, Chapter 8.

Knapp, Robert B. *Social Integration in Urban Communities.* New York, Bureau of Publications, Teachers College, Columbia University, 1960.

Scheinfeldt, Jean. "Opening Doors Wider in Nursing," *American Journal of Nursing,* 67:1461, July, 1967.

Shertzer, Bruce, Ed. *Working with Superior Students.* Chicago, Science Research Associates. 1960.

8

Teaching the Nonprofessional Worker

The principles and methods of clinical instruction discussed thus far are basically those used in teaching all nursing personnel, regardless of how they are categorized. Nevertheless, these methods must be adapted to the work-role and to the educational background of the nonprofessional group—the practical nurse, the nurses' aide, the nursing assistant and the orderly. In this chapter we will consider some methods of adapting techniques of clinical instruction to nonprofessional workers, some of the dilemmas which arise when nonprofessionals are assigned to give nursing care, and some measures which can help the clinical instructor and team leader to work effectively with this group of nursing personnel.

The nonprofessional group is composed of persons with widely different backgrounds and skills, but these workers share certain characteristics which affect clinical teaching. Although they have varying types of preparation, and are assigned to different tasks, most of them function clinically with very limited backgrounds of theory in nursing and sometimes in general education as well. Many are not proficient in verbal skills, such as those required for conference participation. If one believes that all patients should have the care of the professional nurse, who may delegate certain aspects of this care to others, then all categories of nonprofessional workers function as assistants to the professional nurse, rather than as practitioners who assume the responsibility for planning and carrying out nursing care.

Because of such considerations as these, adaptation of methods of clinical teaching will be discussed in relation to the nonprofessional nursing group as a whole, even though this group comprises many different categories of workers.

THE LACK OF EMPHASIS ON CLINICAL INSTRUCTION

Despite mounting concern over the necessity for improvement in the quality of nursing care, the importance of clinical instruction and supervision of nonprofessional personnel is sometimes de-emphasized, even though a large proportion of direct patient care is, in fact, carried out by these workers. Head nurses who are already overburdened with other duties may be expected to provide clinical instruction for student practical nurses, aides, and orderlies. Consequently, aides are sometimes assigned to teach each other, thus minimizing the opportunities for new aides to have the guidance of a professional nurse in acquiring certain nursing skills. Expecting aides to teach each other presupposes teaching skills which many aides do not possess, and it may be the means of spreading misinformation from experienced, but not necessarily expert, workers to new employees. Opportunities for sharing with co-workers is desirable and valuable and should be fostered; this is differrent, however, from planning that the more experienced nonprofessionals will provide the clinical instruction for new employees.

The importance of clinical instruction for the nonprofessional group is sometimes de-emphasized because the administration overestimates the amount of direct supervision which can be given by professional nurses, in light of the other demands which the work situation makes upon them, and also because of misuse of concepts of team nursing.

Two points are particularly important:

1. If the nonprofessional is actually to be an assistant to

the professional nurse, and not merely someone else for her to look after, he must be one on whom the nurse can rely to perform creditably those tasks assigned to him.

2. Nursing is essentially a direct service ordinarily rendered by one worker to one patient; the quality of this service depends on the skill and knowledge of the individual giving the care. Regardless of the category of the worker who changes a patient's colostomy dressing, the skill with which the procedure is performed and the interaction with the patient during the procedure depend upon the skills and understanding brought to the situation by the individual worker.

Sometimes the relative lack of clinical instruction for nonprofessionals is excused on the grounds that only the professional nurse practices nursing. It seems realistic to acknowledge, however, that nonprofessionals do practice some aspects of nursing, and that therefore every effort must be made to help them achieve as high a quality of performance as possible, in relation to the tasks delegated to them. Certainly, present-day nursing requires the professional practitioner to assume responsibility for guiding nonprofessionals in their work. However, it is essential not to confuse this guidance with one's own practice, but to recognize, rather, that it involves helping others to improve *their* practice. For example, conducting a pre-assignment conference does not constitute nursing practice, but is rather a means of helping others to practice more effectively.

Sometimes provision of minimal instruction for nonprofessionals is justified on the basis that these workers will soon be replaced by more highly trained personnel. But since requirements for nursing care continue to grow much faster than the supply of actively practicing professional nurses, nonprofessionals who are assigned to give direct patient care as a "temporary" measure often continue to give this care for many years.

Therefore, rather than minimizing the needs of the nonprofessional group for clinical instruction, we must recognize that their requirements for such teaching are especially urgent, because their direct involvement in patient care is coupled with a limited background of relevant theory upon which to base nursing actions. Nonprofessional personnel, especially, need assistance in making clinical applications of material which is presented by lecture and demonstration. In the absence of this assistance, their performance of such procedures as giving back rubs and bed baths may bear faint resemblance to the method which was demonstrated in the classroom.

Failure to carry out vocational programs effectively does not eliminate such programs, but results in lowering the quality of patient care by denying the nonprofessional workers the opportunities to develop their particular skills as fully as possible. It is essential that the nursing profession support careful preparation of these workers, as long as they give patient care. The task of teaching nonprofessionals with a wide range of abilities presents a challenge to the teacher who must adapt methods so that concepts are presented simply, yet accurately, and so that maximum use is made of the relatively short periods of time available for instruction of this group of personnel.

The Problem of Low Esteem

Our society places heavy emphasis on formal education as a means of social mobility, and tends to look down on relatively unskilled service occupations, particularly if the work involves such "unclean" tasks as emptying garbage or disposing of excreta. Everett Hughes has stated, in *Men and Their Work*, that those who perform lowly or unclean tasks are absolved from the potential uncleanness of these tasks if they are recognized as being among the "miracle workers." (Hughes, Everett C. *Men and Their Work*. Glencoe, Ill., Free Press, 1958.) Thus, the surgeon may become blood-spattered without loss of esteem, because of the recognition of the value of his work, and its dramatic power of healing. The aide, in contrast, typically is not recognized as being among the "miracle workers" and therefore is more vulnerable to being looked down upon by society for performing such lowly and unclean tasks as toileting patients.

The vexing conflict between what is necessary (direct patient care by nonprofessionals) and what is desirable (direct patient care by technical and professional nurses) frequently leads to the nonprofessionals' being accorded low esteem by nurses and physicians, and by patients and their families. Furthermore, because of the shortage of professional nurses in active practice, nonprofessionals are often assigned to perform tasks which are beyond their preparation and competence. In such instances, nonprofessionals function as inadequate substitutes for the professional nurse, rather than as assistants to her, and the problem of low esteem is compounded, because patients, families, and health professionals tend to blame the worker for having inadequate knowledge or skill, despite the fact that such competence is beyond what should be expected. Such situations foster the worker's own sense of inadequacy in performing the work, and may also lead to frustration and anger with patients, as well as with the work situation in general.

Often there is considerable discrepancy in the way the professional and the nonprofessional view the value of the latter's services. The nursing assistant who receives a 6-month training course at a hospital may be the first in his family to receive any kind of vocational training, and may place a prestige value on it. If so, he is likely to be bewildered to discover that his work is considered lowly by the professional staff. However, nonprofessionals tend gradually to reflect the attitudes of health professionals toward the value of the work they perform. Because many nonprofessionals come from deprived social and economic groups, they may be especially sensitive to attitudes of disrespect and condescension which seem but a continuation of the way in which they have been treated by the larger society.

Nonacceptance by the professional staff may lead nonprofessionals to defensiveness, and reluctance to seek the guidance of professional nurses. For instance, they may say, "We're the ones who are with the patients the most. The R.N.'s are too busy giving medicines. What can they tell us about patient care?" The cliché about "being with the patient longest" is now being used by the nonprofessional group, just as it has been used for many years by professional nurses in relation to the amount of time they spend with patients, as compared with the amount of time spent by the physician. Such comparisons are spurious, though, unless consideration is also given to the quality of the care given, and to the length of time required to carry out the care which is appropriate to each health worker's role.

Patients and their visitors also often hold nonprofessional workers in low esteem. Some difficulties which aides and orderlies experience in clinical work are hard for professional nurses to understand, until they realize that some patients and visitors feel free to behave differently toward an aide or an orderly than they would toward a professional nurse. For example, a nurse wondered why an aide seemed to avoid a particular patient, until the aide mentioned that the man "made passes" at her. The aide was amazed to discover that the nurse's suggestion of saying to the patient, "I don't want you to do that" rather than pretending to ignore the behavior and becoming more and more embarrassed, actually worked. Often nonprofessional workers are reluctant to talk with the professional nurse about such matters, thus paving the way for misunderstanding of the way they work with patients and precluding opportunities to receive the guidance they need. For instance, one aide consistently avoided answering a patient's call light, leaving it for the professional nurse to respond to the patient's call. One day when the professional nurse was especially busy, she became angry when she saw the aide walk past the patient's call light, and said, "Why is it you never answer Mrs. Blake's light?" The aide answered, "Once I go in her room, I never can get out—she keeps me there doing things for her. First the bed is too high, or too low, and the window must be opened or closed, and on and on." The professional nurse answered the patient's call but, before doing so, she said to the aide, "Later let's talk about Mrs. Blake's care." In the afternoon the nurse and the aide discussed their experiences in caring for Mrs. Blake. They found that her behavior with the aide was quite different from that with the professional nurse. When the latter answered her light, the patient made one or two requests, and the nurse left. When the aide answered, however, the patient made seemingly endless small requests, to which the aide responded with increasing irritation, and then by avoidance. They speculated about possible reasons for the difference in the patient's behavior toward each of them. Perhaps Mrs. Blake thought of the profesional nurse as too busy, but felt more free to behave toward the aide in ways which would keep her in her room. The professional nurse decided that instead of delegating a good deal of Mrs. Blake's care to others, she would give this care herself for the next few days in an effort to learn about some of the patient's concerns. It would have been preferable, of course, for the aide to have discussed her difficulty in working with Mrs. Blake more promptly, and for the professional nurse to have inquired sooner about the aide's avoidance of the patient. Regularly scheduled informal conferences with non-professional workers provide opportunity for this kind of discussion, so that the professional nurse can decide which patients require more of her attention, and can give better guidance to the nonprofessionals.

In many settings, the work of nonprofessionals involves such close personal contact as giving or assisting with baths, toileting, and lifting helpless patients from bed to wheel-chair. Such activities are usually held in low esteem in our society at large, and within health-care institutions as well. Not only are those who carry out these tasks likely to be considered lowly, but the tasks themselves are sometimes viewed as routine, and occasionally even as unworthy of the attention of the more highly trained professionals. Nevertheless, whether the worker is rough or gentle, whether he respects the patient's privacy, whether he gives a bedpan or urinal grudgingly or willingly, and whether he answers the patient's call light promptly, significantly affect the patient's feelings of acceptance and security.

Nonprofessional workers who render these intimate services typically have little respite from direct patient contact; the patient's call upon the worker has a directness from which professional personnel are somewhat protected. In contrast with the physician, who may spend 5-10 minutes daily with his patient, and the professional nurse, who may spend considerable time in such activities as preparing medicines and charting, which take her away from patients, the nonprofessional's interaction with patients may be almost continuous over an 8-hour period. The nature of his work highlights directness of the patient's call and his lack of opportunity to delegate to others the answering of the patient's requests. For example, calls to a physician are often mediated by an answering service or by a secretary, who can say that the doctor is busy, and ask the patient to call at another time. The professional nurse can delegate to the aide the task of answering call lights, or, if the hospital is equipped with an "intercom" system, she may ask the ward secretary to respond to calls.

The answering of call lights is viewed by some persons as the epitome of "being at the beck and call" of others, and is an activity which is frequently delegated to nonprofessional personnel. (One may question, however, whether responding to a patient's call is so simple an activity as is sometimes assumed, requiring, as it does, a high degree of sensitivity to the patient's unspoken, as well as to his spoken requests. Unless nurses are in close contact with patients, the use of the intercom presents further difficulties. To the fearful preoperative patient, the mechanical-sounding "May I help you?" coming from a speaker in the wall may leave him at a loss for words. How does one say, via the intercom, "I'm frightened." It may seem more appropriate to the patient to say, "I'd like an extra blanket"—and for the secretary to ask the aide to take a blanket to the patient.)

SOME WAYS OF DEALING WITH THE PROBLEM OF LOW ESTEEM

The low esteem often accorded to services rendered by nonprofessionals, the close personal nature of these services, the nonprofessional worker's comparative lack of training for these tasks, and the extensive time periods when they have contact with patients, all emphasize the need for more effective education and supervision of these workers, and for measures which provide them some respite from patient contact.

Providing periods away from patients

Providing for periods away from patients may seem utterly unnecessary to the head nurse or supervisor who has spent valuable time hunting for nursing assistants who occasionally seem to vanish into thin air. When workers hide in linen closets, or kitchenettes, their action may stem from the need to protect themselves from continuous demands of the clinical situation. Changing a colostomy dressing for the twelfth time in one day, while still showing acceptance and concern for the patient, is taxing—whether one is a professional or not. The "coffee break" is a common and useful way to provide for rest periods. Rotation of assignments so that no one worker consistently has contact with the most difficult patients, and interspersing cleaning assignments with assignments to patient care are helpful in lessening fatigue and strain. And, although the purpose of conferences and demonstrations is to foster improved patient care by helping the worker to increase his knowledge and skill, they also have a useful "side effect" in providing a change of pace.

Providing help with emotional reactions to patients

Some of the strain of working with sick people can be lessened by learning how to relate to them more effectively, and how to deal with one's own reactions to clinical experiences. Sometimes, however, it is assumed that brief demonstrations of technique will suffice in teaching nonprofessionals, although the professional nurse's education includes consideration of her emotional reactions to giving care, to the meaning of touch in nursing, and so on. Opportunities for discussion of such aspects of giving care (geared to an appropriate level) are necessary for anyone who is to render care. Because nonprofessionals are expected to provide a large proportion of patients' physical care, particular emphasis should be placed, during their training programs, upon discussion of their reactions to such tasks as care of incontinent patients, and to instruction in gentle and respectful ways of handling helpless persons. The teacher's attitude (and that of the other professional nurses on the ward) toward patients is especially important in influencing the way the nonprofessional worker relates to patients, because he is usually quick to reflect the attitudes of the professional staff. Recognizing the need for such instruction does not imply that nonprofessionals are lacking in human warmth and feeling, but that they require assistance in learning how to channel these qualities to the benefit of their patients. For example, an eighty-six-year-old patient who suffered from frequency and urgency got out of bed during the night and voided on the floor, in order to avoid soiling her bed. The aide scolded the patient, returned her brusquely to bed, and noisily replaced the siderails, instead of putting a bedpan within the patient's reach and supplying a call bell. Later the aide said she felt sorry about the old lady's predicament, but that the puddle of urine on the floor so disturbed her that she punished the patient, instead of helping her.

Providing recognition of the value
of nonprofessional services

Measures which help the worker recognize the value of his work, and which convey respect for the worth of his services can make it easier for him to seek guidance from professional nurses, develop greater interest in giving care to patients, and derive more satisfaction from his work. Provision of attractive uniforms and calling the worker Mr. or Mrs., rather than John or Mary, are obvious measures, as are careful selection of applicants for the job, and provision of adequate salaries. The idea that "anyone can get the job" is an important factor in lowering esteem; it also fosters the vicious circle of low pay, poor selection of applicants, inadequate training, and retention of workers whose performance is substandard.

The values which the professional staff place upon the tasks assigned to nonprofessionals are crucial in determining how effectively these tasks will be carried out. Those who are responsible for instruction and supervision of nonprofessionals must value the importance of such everyday measures as keeping the Sitz bath clean, or of assisting patients to take tub baths or showers. If the professional nurses believe that the more technically complicated a procedure the greater its value, they will communicate this attitude to the nonprofessionals, and undermine the workers' interest and motivation for the tasks to which they are assigned. During conferences, professional nurses may imply that what is really significant—the use of various types of respirators, for example—is beyond the areas in which nonprofessionals function. Others may attempt to teach nonprofessionals things which ordinarily lie outside the sphere of their activities, in an effort to make their work more interesting (according to the nurses' values.) This practice is similar to that of some physicians who offer to instruct professional nurses in differentiating various types of heart murmurs, as a gesture toward according her more status and making her work more interesting.

Providing clarity about the nonprofessional worker's role

The differences between the nonprofessionals' role, and that of the professional nurse, need to be clarified. Otherwise, the problem of low esteem may be heightened by the workers' belief that they do almost the same work as the professional nurses but receive less pay and prestige. Such differentiation can be made in a matter-of-fact manner during class or conference without implying that one category of personnel is more useful or worthy than the other, but that each performs a specific role. For instance, in one conference about care of diabetic patients the professional nurse stressed the importance of giving meticulous foot care, testing urine accurately for sugar, and noting whether any food was left on the patient's tray. However, she also briefly stated the professional nurse's additional responsibilities for such matters as teaching diabetic patients and their families. Mention of the professional nurse's role not only helps the nonprofessional group to understand the differences between their responsibility and education and hers, but also assists them to develop a broader view of the plan for patient care, and of the part they play in implementing this plan.

Often the nonprofessional worker needs help in learning to differentiate between a work role and a social role in his relationships with patients and staff. During the education of professional nurses considerable emphasis is placed upon behavior appropriate to certain roles, but this aspect of instruction is frequently not emphasized sufficiently during the training of nonprofessionals. The need for such instruction is acute, however, as all of us realize when we hear an aide, for example, relating her marital troubles to a patient, or when we find that an orderly expects professional staff to spend a good deal of time advising him about such personal problems as disciplining his children. Too often the workers' confusion of roles is handled solely on a disciplinary basis. Signs may be posted in the ward kitchen and utility room, for instance, warning against such infraction of rules as having a cigarette with a patient. Although discussion of roles must be kept simple, much can be accomplished by using everyday examples of the differences between behavior expected in the social role and in the work role. For example, one middle-aged woman recently employed as an aide had never worked outside her home before. When she served breakfast to her patients, she saw nothing wrong with making a cup of coffee for herself and joining the ward patients as they had breakfast. She was both puzzled and hurt by a reprimand for her behavior, but a simple explanation about the difference in her role when serving coffee in her home and serving it to patients in the hospital served to clarify the matter.

TEACHING AND SUPERVISING THE NONPROFESSIONAL

Instruction by teachers

The process of teaching the nonprofessional can demonstrate respect for the worth of his work. When instruction of the nonprofessional is taken seriously it implies that what he does is important, and that the way it is done matters. When the nonprofessional commences his clinical work, instruction and supervision should be carried out by teachers; initial instruction is thus handled by persons who are not so burdened with other responsibilities that they have little time for teaching the new nonprofessional worker.

Because these workers function with limited backgrounds of theory in nursing, instruction should be concrete and highly specific. For instance, the appropriate temperature for a Sitz bath should be stressed and a thermometer provided for testing the temperature of the water. An approach to teaching which conveys "This is how it is done. If it is necessary to change this procedure in any way, the professional nurse will explain what change is needed, and help you to make it" is more effective with this group of workers than one which points out varieties of alternatives, depending on the requirements of different situations. With increasing education of professional nurses, and greater emphasis upon the nurse's using her own judgment in adapting procedures to the needs of patients, it is essential to recognize that this approach is not applicable to workers with limited grasp of theory. Sometimes emphasis on correctness of procedure is viewed as "old fashioned" and "rigid," regardless of the category of personnel being taught. However, "flexibility" and "emphasis on principles" can turn into laxness and hazardous practices in the hands of persons who are not prepared to use the latitude which has been accorded them.

Discussions between professional nurses and the instructor concerning the preparation which each of the nonprofessional workers has had, and his job description, are essential, so that assignments can be planned in accordance with each worker's preparation. During the training period the worker's educational requirements must be uppermost in planning his clinical assignments. For example, the student of practical nursing is no less in need of carefully planned and supervised learning experiences than the student of professional nursing. The teacher and the professional nurses on the ward should collaborate in planning these experiences.

Some of the approaches which the teacher uses in guiding the nonprofessional can be continued later by the professional nurses, after the worker has completed his training period. For example, conferences concerning patient care are initially conducted by the teacher. After the training period, the nonprofessional should attend conferences conducted by professional nurses on the ward.

Clarification of goals and terminology

Before beginning any clinical teaching, it is important to state one's goals clearly. Sometimes lack of clarity concerning goals leads to approaches to teaching which convey that the nonprofessional is being given a brief course in professional nursing. Such an approach can lead only to confusion for the worker, and frustration for the teacher, since she cannot possibly provide a facsimile of the professional nurse's education in the limited time available for teaching nonprofessionals. Although clarity of terminology is only the beginning, it is a necessary step in defining the role and the type of instruction being planned for the non-professional. A course for aides in the care of children, for example, should not be titled "Pediatric Nursing" or "Nursing of Children," but something like "Assisting with Child Care." When the nonprofessional receives a clinical assignment, the words spoken should convey that he is an assistant to the professional nurse, rather than implying that he is being given responsibility for the nursing care of several patients. "Mr. Brown, today I want you to help me care for Mr. Walsh, Mr. Jones, and Mr. Leary," is preferable to, "Mr. Brown, you will have Mr. Walsh, Mr. Jones, and Mr. Leary today."

Although viewing the nonprofessional as an assistant to the professional nurse is the essence of team nursing, this concept is sometimes distorted when team nursing is carried out. The impression may be given during team conferences that "We do everything together," and that every patient care problem will be handled by the group, thus conveying the idea that there is little difference in the roles and responsibilities of the various team members. "How can we teach Mrs. Smith about her diet?" may lead to a discussion which does not bring the expected results in the patient's learning, because members of the group do not clarify what their roles are in patient teaching, and how well prepared each is to carry out teaching. Is the aide, who understands little about diabetic diets, able to teach Mrs. Smith about her diet? Would it not be preferable for the professional nurse to indicate, during the conference, that she will work with the dietitian in helping the patient learn about her diet, but that, because such instruction is important for this patient, nonprofessionals should be alert to Mrs. Smith's questions and reactions to her diet, and to how well she eats, and report these observations to the professional nurse?

Guidance in communicating with patients

Because nonprofessionals spend so much time with patients it is essential to stress certain simple concepts to guide them in communicating with patients. Many training programs for nonprofessionals place little emphasis upon ways of talking with patients. However, each time the worker is with a patient he is communicating with him in some fashion, and he requires guidance so that this interaction can be helpful. At the very least it should be innocuous, and not constitute a source of annoyance or worry for the patient. Many nonprofessionals can, with guidance, become sympathetic listeners and alert reporters (to the professional nurse) of patients' questions and concerns.

Discussions with nonprofessionals should stress such basic guidelines as not recounting one's personal affairs to patients, and noticing the questions and concerns mentioned by the patient. Too often this resource for learning about patients' concerns is under-utilized, because nonprofessionals are not adequately instructed in the importance of listening carefully to what patients say. For example, an elderly woman may wonder aloud to an aide where she will go after leaving the hospital, since she has no one at home to care for her. While such a comment would alert the professional nurse to the patient's need for help in planning her future care, an aide may pay little attention to it, and say nothing to the professional nurse about it. By repeated discussion of samples of patients' conversation with nonprofessionals, these workers can become more aware of the kinds of communication from patients that warrant being reported to the professional nurse.

Other basic concepts concerning relationships with patients include avoiding a "tit-for-tat" attitude, moralizing, pressing discussion of religion upon patients, and discussing one's own health problems with them. If instruction about such matters is presented only in lecture form it can readily seem like a list of "do's and dont's," but it is quite effective when related to patient-care situations brought up by the workers. For instance, one practical nurse recounted her experience with a patient who was irritable and who spoke to her crossly. Her approach had been to match his irritability with her own, thus creating a situation in which patient and nurse were cross with each other. She justified her behavior by saying, "I don't take that from anybody." It was several weeks before she began to realize that her approach made her patient contacts unpleasant not only for them, but for herself as well, and it was then that she began to modify her behavior.

Role-playing is a teaching device that can help make some concepts concerning conversing with patients more meaningful. Often it is helpful to re-enact a conversation between the worker and patient during a group conference, as a basis for discussing other ways in which the situation might have been handled. For example, a patient who made many requests said to an aide, "I know I'm a nuisance," to which the aide replied, "I'm glad you said that; I didn't." Although the aide's reply sounds callous, she was puzzled about how to answer the patient, and welcomed the opportunity to discuss other possible ways of replying to the patient.

Adaptation of instruction to individual
capability and role

Nonprofessionals are likely to have very diverse backgrounds, and to vary considerably in their intelligence. While some are limited in their abilities, others demonstrate keen intelligence and initiative, but are prevented from entering a professional program by such circumstances as lack of money or family responsibilities. Married women who live in areas where job opportunities are not plentiful or varied often seek employment as aides in a nearby hospital. These women frequently bring intelligence, maturity, and dependability to a type of work which in urban areas is more likely to be performed by persons of lesser capability and aptitude.

Considerable adaptation of teaching is required in light of differences in individual capability. Such adaptation must be made within the framework of the goals of the program, however. For example, the terms used in a conference with several very intelligent aides who have had previous contact with medical words can be more sophisticated than those used with aides of very limited educational background and intelligence. In neither case, however, should material be presented which would prepare the worker for tasks beyond those he is permitted to perform in his particular setting.

When assessing the backgrounds of nonprofessionals, it is important not to assume that these workers necessarily have less to offer in all aspects of patient care than the professional nurse. The diversity of backgrounds, intelligence, and initiative among these workers sometimes leads them to make important suggestions and to report observations which have escaped the attention of the professional staff. One student practical nurse (who was Hungarian) was informed that a certain psychiatric patient was mute. The student noted that the patient's name was Hungarian. After addressing the patient in English and receiving no reply, she spoke to the patient in her native language, whereupon the patient's "muteness" disappeared.

Group conferences to discuss patient care, demonstrations, individual conferences, and carefully supervised clinical practice must be adapted to the level of the group's ability and work role.

When leading conferences one must remember that the nonprofessional's verbal skills are often limited. Participating in conferences may be a new experience for some members of the group, and they may be not only awkward in expressing themselves, but reluctant to do so. Thus it is important for the leader to use simple terminology, and to ascertain what the participants understand about the topic before proceeding with the discussion. "Tell me what you know about ......" is a good way to begin discussion of a topic. Lack of verbal facility and of medical knowledge should not be equated with lack of intelligence or of interest. Content must be carefully adapted to the individual's role;

the teacher must resist the temptation to try to make the nonprofessional worker into the nearest possible likeness of a professional nurse. For example, discussion of the care of a patient with emphysema should focus on such topics as the need for providing rest periods during the morning bath so that the patient does not become fatigued and more dyspneic as a result of the procedure, and the importance of answering his call light promptly, rather than on the use of various types of respirators, the dynamics of pulmonary malfunction, and the physiologic effects of various medications.

Although it is important to concentrate on the "doing" it is important not to leave out the "why." It is as true for the nonprofessional as for the professional worker that motivation to do a task correctly is increased when one understands the reasons why it should be done in a certain way. Thus, reasons for certain actions should be stated simply and concretely. Instead of saying, "Do not wrap the Ace bandage too tightly" it is preferable to say, "Wrapping the Ace bandage too tightly could interfere with flow of blood to the foot."

It is the professional nurse's responsibility to foster discussion during conferences, and to see that it is related specifically to the workers' experiences with patients, rather than using the time to present theoretical material without stressing the application to the care of patients. When discussing such an incident as a fight between two psychiatric patients, a lecture on sibling rivalry and its implications would be less effective with a group of attendants than a discussion of how they actually handled the situation, and of other measures which might have been preferable. It is important to give nonprofessional workers an opportunity to say how they actually handled nursing situations. One aide told how she cared for an elderly man who asked frequently for the bedpan but was unable to defecate when placed on the pan. When asked how she handled the situation the aide replied, "I just told him to go ahead and go

in the bed, because I'd have to change the bed later anyway."
The discussion which ensued brought out the reasons why
the action that appeared to the aide to be a logical time-
saver was, in fact, undesirable for the patient.

When evaluating information concerning the patient's
diagnosis and personal life which is to be shared with the
nonprofessional group, three important considerations are:

• Do the nonprofessionals need this information in order
to give adequate care?

• Are patients' interests served by sharing this information with persons outside the professional group?

• Are the nonprofessional workers likely to understand
this information and be able to use it constructively?

Sometimes information is imparted very freely, in the
belief that everyone who works with the patient should
share equally in information concerning him. Where a
large group of people with varied backgrounds share information about a patient the problem of maintaining confidentiality becomes acute. Diagnostic terms may be used
freely, but are not explained sufficiently, and this leads to
misunderstandings. For instance, one aide thought the diagnosis "threatened abortion" meant that the patient was
threatening to rid herself of the child; the aide consequently
avoided the patient. Diagnoses such as syphilis present particular problems. Although the patient is not infectious, the
aide may avoid going near him for fear of catching the
disease. In this instance, for example, the patient's care may
be facilitated and confidentiality fostered if the patient's
condition is described as "an infection being treated with
antibiotics." If the nurse knows that a patient who was
admitted with burns acquired them by falling over a hot
stove while intoxicated, what purpose is served by relating
this information during team conference? Would it not be
preferable for the professional staff to make the necessary
observations related to the patient's history of overuse of
alcohol and, as long as this problem does not come up during

the patient's hospital stay, to allow this information to remain with them, rather than discussing it with the non-professional workers?

Whether the experience is a clinical conference, a formal class, or a demonstration, it is important not to attempt to include too much detailed information in one session. People who are not accustomed to attending conferences and classes sometimes find it difficult to concentrate on educational activities for extended periods. Some nonpro-fessionals, especially if they are older and have been away from school for a time, may resist taking part in such educational activities as conferences and demonstrations. Often they do not think of themselves as students, and when they sought this type of employment, did not anticipate the necessity for such educational experiences. School experiences have been unrewarding for some of these workers, and they tend to avoid contact with educational activities. However, as they discover that their observations and contributions are welcome, and are acted upon, many nonprofessional workers enjoy participating actively in classes and conferences.

Because of these workers' limited verbal skills and their restricted knowledge of theory, it is important to use demonstrations liberally when teaching them. Showing the aide how to feed a patient is much more effective than just discussing such points as not rushing the patient, allowing him to assist if he is able, and so on. Opportunity for the aide to re-demonstrate is important too. The worker who is insecure is all too likely to reply affirmatively when asked if he knows how to perform a procedure. Going with him to see if he needs assistance, or, if this is impossible, asking him to describe how he will proceed, is preferable to asking, "Do you know how?" to which the worker is likely to reply, "Yes" whether he knows how to proceed or not.

The range of tasks a nonprofessional worker is expected to perform should be limited sufficiently so that he masters the material he is taught, and shows himself competent in applying it.

Discussion of legal responsibility

It is important for the teacher to discuss with the professional staff nurses their individual and specific legal responsibilities for supervising the nonprofessional workers. Also, it is necessary that the nonprofessional workers be instructed concerning their own legal responsibilities. Too often both groups have misconceptions about this matter. Nursing assistants sometimes share with professional nurses the misconception that the professional nurse assumes legal responsibility for the clinical work performed by nonprofessionals. It is not unusual for an aide to shrug and say, "Well, the nurse is responsible for what I do, I'm not." It is important for both nonprofessional workers and professional nurses to recognize that each individual is responsible for his own actions, and that although the professional nurse is accountable for carrying out her responsibility for supervision prudently, this does not relieve others of responsibility for their own clinical performance. Stressing this point can help nonprofessionals to recognize the importance of seeking help when they are not sure how to proceed, and refusing to carry out tasks which are beyond their preparation and competence. It can also emphasize to the professional nurse her responsibility for planning assignments which are within each worker's job description, and for providing effective supervision for nonprofessional workers.

ORIENTING THE NONPROFESSIONAL TO HIS CLINICAL ASSIGNMENT

The orientation of the nonprofessional to his clinical assignment should convey that he is expected to assist with care, rather than to take on the responsibility for any patient's nursing care. He should be instructed as to the specific aspects of care which he is to perform, and advised which professional nurse is responsible for the patient's care, so that he may consult with her about questions and problems which arise. Often listening to the "morning report" with all other nursing personnel is the only orientation the nonprofessional worker has to his assignment. Although it may be helpful for him to attend these reports because it helps him to feel part of the group, and gives him opportunity to learn about the patients with whose care he will assist, such attendance is no substitute for orientation of the worker to his assignment. He requires a separate briefing, which brings out in simple terms the care he is expected to give each patient, the observations he is expected to make, and so on. The nonprofessionals' ability to make the necessary connections between what he hears during the nursing report and what he is expected to do for patients is usually limited, and unless he has a separate briefing which is geared to his understanding and to his role, he may not realize, for example, that the night nurse's brief comment, "He is a G.I. bleeder" means that he should save the patient's stools and vomitus for the professional nurse's inspection.

After the morning report and orientation to their assignments, the nonprofessional workers should be introduced to patients by the teacher or by the team leader. This can convey to the patient that the aspects of his care that are to be carried out by the nonprofessional are guided by the professional nurse, and that she has confidence in his ability to perform them. If this introduction includes an explanation of what the worker will do for him, the patient can be spared worry over whether the student practical nurse is qualified to give him his enema, for example. When the patient has confidence in the nonprofessional worker's ability, the relationship between them is facilitated and the patient is more likely to view the nonprofessional as an assistant to the professional nurse, rather than as an inadequate substitute for her.

ASSISTING WITH AND SUPERVISING THE CARE GIVEN
BY THE NONPROFESSIONAL

It is desirable that the teacher and the professional
staff nurse frequently participate with the nonprofessional
worker in carrying out his nursing care assignments. Initially this is needed to guide him in learning how to give care,
and later in providing opportunity for him to continue
learning how to adapt the nursing care so that it meets
various patients' requirements.

The nurses' aide, particularly, requires the assistance of
the professional nurse in making adaptations in patient
care. Unless such guidance is given, care becomes routine
because of the limited ability of nonprofessionals to assess
nursing requirements of patients. For instance, the aide may
bathe all patients if it has not been brought to her attention
that some of them can assist with their own baths. She may
also require specific directions concerning which patients
should be cared for first; otherwise there is a tendency for
the nonprofessional to go first to the patient whose name
appears first on her list. It is also important for the professional nurse to be present when certain aspects of care are
given, so that she can observe how the worker performs
various tasks, and what kind of assistance or instruction
she requires.

Whenever possible, nonprofessionals should be assigned
to assist with care of only those patients whose requirements
and reactions to care are known to the professional staff.
Merely repeating such general instructions as "Encourage
the patient to help himself" or "Encourage him to take
fluids" are meaningless if a worker does not know how to
implement them with a particular patient. The instruction,
"Gain the patient's confidence, so he will let you care for
him," would not be particularly effective. When a patient
in traction cried out, "Oh, my leg," even before anyone
approached his bed, he was cared for by the professional
nurse in order to ascertain what nursing actions could lessen

his apprehension and enable him to accept care from others. In this instance, the nurse found that a firm approach, showing that she knew just how to move him, along with touches of humor, were effective in lessening the patient's apprehension. After she had bathed the patient several times, she asked the practical nurse to assist her with the care, thus giving this worker opportunity to observe how the patient was moved, and the approach which the professional nurse used with the patient. The following day the practical nurse gave the patient his bath and managed it very well.

Nonprofessionals require opportunities to derive satisfaction from their work. Too often such satisfaction is limited because the worker is frequently interrupted by having to run errands. Or, his assignment is changed so frequently that he may have little continuity with patients. His questions about patient care may go unanswered (and sometimes unasked) and he may have few opportunities to discuss his emotional reactions to giving patient care. Boredom and resentment often result. Improved planning can result in fewer interruptions for all categories of nursing personnel. For instance, if one worker is asked to be the messenger for the day, taking patients to other departments, running errands, and so on, the rest of the staff will be able to continue giving care with relatively few interruptions. When the assignment to act as messenger is rotated, no one worker experiences the boredom which doing this kind of work usually produces. Careful planning of clinical assignments can also provide the nonprofessional with continuity of contact with patients, thus permitting him to give more personalized care and to learn more about the patients he cares for.

Regular individual conferences are needed, too, to discuss how the worker is performing, and to make suggestions concerning his work. The nonprofessional group often dread such conferences at first, until they find that these periods help them to gain greater satisfaction from their work. Because many of these workers are very apprehensive about such conferences, it is important to proceed slowly in offering criticism and suggestions.

One of the most pressing requirements for clinical instruction of nonprofessional workers is simple, readily available reference materials. *The Handbook for Nursing Aides in Hospitals* and the more recent publication, *Training the Nursing Aide Student Manual* (both available from Hospital Research and Educational Trust, Chicago), are examples of useful publications. Although most hospital procedure manuals are not readily understood by the nursing assistant group, many of these workers can benefit from reviewing simply written instructions, such as those for giving a partial bath, making a bed, or testing urine for sugar, and which are provided in manuals prepared specifically for nonprofessionals. By providing necessary reference materials the professional staff are saved time in answering such questions as "What is the correct temperature for a hot water bottle?" and nursing assistants' initative in seeking information can be fostered.

JOINT INSERVICE PROGRAMS FOR PROFESSIONAL
AND NONPROFESSIONAL WORKERS

Having professional and nonprofessional staff attend the same inservice lectures or demonstrations is common practice. Although it presents opportunities for learning, it has many pitfalls. Mixing professional and nonprofessional groups at lectures and conferences may result in a level of discussion which is inappropriate for both groups. It may also tend to confuse the roles of the two types of workers, unless the differences in roles are made very clear during the conference. Sometimes mixing the groups is advocated because of the mistaken idea that this is an expression of democracy in administration. Another reason for having joint inservice programs is the belief that, even though the content is not planned for the nonprofessionals, they may glean something from the discussion, when time does not permit having separate programs for each group. If nonprofessionals are asked to attend educational programs with the professional group, separate discussion should be held with them later to clarify points which they do not understand, and to help them recognize the application of the content to their own work with patients. This would obviate, for example, such a puzzled reaction as that of a practical nurse who attended a physician's lecture on the technique of closed-chest cardiac massage, and who remarked to a friend as she was leaving the lecture hall, "I wonder if we are expected to do that, or if only the R.N.'s are supposed to do it."

Fostering quality care, despite dearth of professional nurses
    Assigning nonprofessionals to give patient care presents the possibility of a compromise with the quality of care, regardless of how simple the procedure, or how well the nonprofessional performs it. If the professional nurse were giving the care, she could make on-the-spot adaptations in the care as indicated by the patient's condition, and she could make more astute observations of his responses. For instance, the practical nurse may not notice the puzzled look on a patient's face while she is taking his temperature, and she may therefore not give the patient the opportunity to voice his bewilderment, or report the patient's reaction to the professional nurse. The orderly who helps a man with a Sitz bath may not respond effectively to the patient's question of how he should check the temperature of the water when he carries out the procedure at home. The person who is with the patient is the one who hears his questions, and has opportunity to notice his responses to care. The more contact the professional nurse has with the patient, the less the problem, since she has many opportunities to observe the patient. This is not always possible, however, particularly in such settings as nursing homes, where a large proportion of care is given by nonprofessionals. In these situations, particularly, the professional nurse should stress the

importance of nonprofessionals' reporting to her not only patients' physical symptoms, but the questions which they raise, and their reactions to care. Discussion of these points in individual and group conferences can increase the non-professionals' awareness of the significance of observations of patients' behavior and concerns. During conferences the professional nurse should ask not only such questions as, "How much did the patient void?" but "What questions did the patient ask?" and "What did the patient talk about?" Although learning of patients' concerns in this way is not as effective as learning by frequent interaction with the patient, it can alert the professional nurse to some of the problems which patients are experiencing, so that she can spend more time with certain patients, as well as guide the nonprofessional workers in caring for them.

Conclusion

As long as a large proportion of direct nursing care continues to be given by the nonprofessional group, it behooves us to make every effort to insure that the care they give is of the highest possible quality. By acknowledging the importance of the contribution made by nonprofessionals to patient care, showing respect for them, and providing them with the necessary instruction and supervision, we can accomplish a great deal toward improving the quantity and quality of nursing service available to patients.

SUGGESTED READING

Edelson, Ruth E. "A Retraining Project for Preparing Men Practical Nurses," *Nursing Outlook, 14*:33, August, 1966.
Frye, Lillian B. "An On-Duty Inservice Experiment for Aides," *Nursing Outlook, 13*:60, August, 1965.
Hall, Madelyn N. "Home Health Aide Services Are Here to Stay," *Nursing Outlook, 14*:44, June, 1966.
Hughes, Everett C. *Men and Their Work.* Glencoe, Ill., Free Press, 1958.
Mansfield, Elaine. "Use of Patient Care Plans by Aides," *Nursing Outlook, 15*:72, April, 1967.

Rasmussen, Etta H. "Preparation of Faculty for Schools of Practical Nursing," *Nursing Outlook, 13*:52, October, 1965.

Rasmussen, Sandra. "Medicare and the Licensed Practical Nurse," *Nursing Outlook, 14*:62, June, 1966.

U.S. Department of Health, Education, and Welfare, Public Health Service. *Nursing Aide Instructor's Guide.* Washington, D.C., United States Government Printing Office, 1953.

9

Working with Clinical Agencies

The role of the faculty member, and her working relationship with personnel of service agencies, vary in different programs and reflect the beliefs of faculty and educational administrators about what this role should be. One belief holds that, in order for a program of clinical teaching to be effective, the faculty must have some control over the conditions under which nursing service is rendered. Such control usually entails a dual position, in which one individual carries responsibility for both nursing education and nursing service.

A different belief concerning the role of faculty holds that faculty should be involved only with teaching, and that they should not also carry responsibility for nursing service. When this view is taken, faculty and administrators arrange for students' clinical work to be done in a variety of community health facilities which may or may not be affiliated with the university. Thus, faculty may go to several different agencies, according to the clinical-experience requirements of their students.

These divergent views concerning the appropriate role of faculty are evident in both collegiate and diploma schools. The faculty of some diploma schools and some collegiate schools carry responsibility only for education of students; in other diploma and collegiate programs the faculty have a dual role in service and education.

Both of these arrangements have advantages and disadvantages for the teacher as she works with students and patients. The individual with a dual position has a considerable amount of control over the standards and practices of nursing care of all patients in the areas for which she is responsible. She is in a position to effect change in the nursing service situation. She must also, however, cope with the conflicting demands which can arise from her dual role. For instance, she may find it necessary to cancel scheduled conferences with students because of pressing nursing service demands. She may discover that nursing service requirements do not allow her the time for study and planning which are necessary for effective teaching. In many situations, while her control of the over-all nursing service situation is considerable, she has little opportunity to become involved in the direct care of patients because of the pressures of her administrative duties. In some instances, however, the person with dual responsibility functions in a situation which does permit her to spend a good deal of time in direct care of patients, as well as in bedside teaching. Under such circumstances she is in a position to provide expert care and clinical instruction.

The faculty member who is responsible only for education may feel frustrated because she has little control over standards and practices of nursing care in the institution. She may also be at a disadvantage when she tries to help students work with such agency personnel as staff nurses, physicians, and dietitians, because she herself has fewer opportunities to work with them. It may be difficult for her to interpret to students the nursing service problems of the agency because she does not deal directly with these problems. On the other hand, it may be easier for her than it is for the teacher in a dual position to view some clinical problems objectively, because she is not a member of the agency staff. Also, she is less likely, perhaps, to interpret a student's question concerning patient care as criticism of her, of the staff, or the institution, and she may therefore be better able to clarify the issue with the student in a way which fosters learning. Then too, she may be more likely to draw students' attention to approaches to nursing care problems which are used in other clinical settings, but are not in common use at the agency where students are practicing.

Although it is ordinarily easier for the teacher who is responsible solely for education to use such measures than it is for one who has a dual position, freedom to examine various approaches and ideas concerning nursing is essential for every teacher of nurses. In the past, nurse instructors have too often had this freedom abridged, sometimes by hospital administrators, sometimes by physicians; they were therefore hampered in fostering the breadth and flexibility of inquiry so essential to liberal education.

Regardless of whether or not the teacher functions in a dual role, it is important for her to consider the advantages and disadvantages inherent in her particular arrangement, and to make an effort to compensate for the disadvantages. The teacher who carries no responsibility for nursing service can help compensate for her lack of influence on the over-all quality of patient care, and her lessened orientation to the problems of the service agency, by suggesting joint conferences of staff, faculty, and students for discussion of patient care and nursing service problems. Students and faculty thus have greater opportunity to contribute to improvement of patient care, and to share ideas with nursing service personnel.

It is essential that the nurse faculty member be a practitioner of nursing, whether she does or does not have a dual position. In addition to observing the care which students give to patients, she should, herself, give some direct nursing care in the presence of students, thus providing them with a role model. This will also permit her to acquire information about the patient through her own observation and participation which enables her to guide students' learning more concretely and effectively. Any teacher who works directly with patients and students has a dual role, in the sense of concern for and responsibility toward both patient and student, regardless of whether she carries responsibility for nursing service. The patient benefits from the teacher's skill and judgment when she provides some of his care, and also when the student provides care with the teacher's guidance.

By participating directly in nursing care, the teacher also indicates to the staff her interest and competence in nursing. Her participation takes many forms: regulating the flow of an intravenous infusion; reporting significant symptoms; seeking and sharing with the staff additional information about the patient. Such participation provides a sound basis for working with agency staff, because it focuses on mutual concern for the patient, and on collaborative effort to provide nursing care. It also demonstrates respect for the practice of nursing. The teacher who is preparing practitioners of nursing must demonstrate by her own actions and attitudes that she respects the work of the nurse practitioner. Subtly deprecating the practice of nursing (by overemphasizing the *knowing* and underemphasizing the *doing* aspects, for example,) can infect students with doubt about the value of the work for which they are preparing, and can seriously impair working relationships with the nursing staff.

In addition to maintaining concern for and participation in the practice of nursing, at least three other considerations are useful in creating an experience within the clinical agency which will be fruitful to the agency itself, the instructor, and the student nurse: 1) planning ahead for the agency experience; 2) developing an effective plan for orienting the students to the agency and the agency to the students; and 3) establishing constructive working relationships with agency personnel.

PLANNING AHEAD FOR THE AGENCY EXPERIENCE

Whether faculty and students are guests of an agency, or function in a "home" hospital affects the planning for clinical experience. If an educational and a service agency are involved, personnel of both institutions must discuss the facilities and equipment needed in order to carry out the teaching program, and decide which institution will provide these facilities, or whether they will share in providing them. Pressures for space and equipment are so great in most universities and hospitals that unless definite arrangements are made, the teacher may find that neither institution has assumed responsibility for providing such needed facilities and equipment as lockers for faculty and students, a filing cabinet or desk space in which the teacher may keep instructional materials, and use of conference rooms and the library at the agency.

Requests for particular experiences and facilities must usually be more specific when faculty and students are guests at an institution. In a "home" hospital it may be taken for granted that a student may have observational experiences in the operating room, or obtain charts from the record room, but such activities ordinarily require administrative approval when one is a guest in the institution.

Finally, any rules of the host agency that differ from those of the home institution should be carefully ascertained in advance and explained early in the students' orientation to the agency.

Developing an Effective Orientation Program

Planning for effective orientation to the agency is especially important in programs where faculty and students must move to different clinical settings in order to obtain necessary clinical experiences.

Rules of the agency must be observed while students and faculty are practicing there. Although this may seem obvious, confusion sometimes arises when the two institutions have different sets of rules. For instance, it may be permissible to smoke in college classrooms but not in hospital conference rooms, or vice versa. If a hospital rule states that no patient may have visitors who are under the age of twelve, the teacher is not free to suggest that a child be permitted to visit unless she first finds out whether an exception can be made to the rule. Observing the rules does not necessarily mean that a teacher personally agrees with them, or that she cannot discuss with students alternate ways of functioning which might be feasible in another setting.

Faculty must find ways of adapting quickly to new clinical environments, and of helping students to do so. It is important to differentiate between the information required by faculty and students, and that required by agency personnel. For example, it is ordinarily not necessary for the student and the teacher to make out the variety of forms and requisitions used in an institution in order to perform their roles in teaching and learning how to give nursing care. Information about such forms should therefore be included in the orientation of staff nurses to the work setting, rather than in the orientation of students and teachers to the agency.

Orientation to the agency not only benefits students and faculty, but it also protects patients and agency personnel from confusion and error which can arise when groups of people (of whatever category) begin to function without sufficient knowledge of institutional policies and practices. It is essential for agency personnel and faculty to work together to develop an orientation program which is designed to:

• Assist students and faculty to function with maximum ease and safety in a new setting.

• Shield staff from unnecessary and repetitive demands upon their time and energy in answering questions and locating supplies.

• Safeguard patients from care given by persons not sufficiently oriented to their nursing requirements.

Such orientation programs invariably involve compromises between what is ideal and what is feasible in light of the time available. One might, for instance, think it would be ideal for students to spend an entire week or even longer in becoming oriented to the agency, before any clinical practice is begun, but this would seldom be feasible due to pressures to provide necessary clinical experiences within the limited time allotted for clinical practice.

The amount of time spent by faculty and students in becoming oriented to new settings must be considered. Each change in clinical environment necessitates expenditure of time and effort on orientation—time and effort which could otherwise be spent in the ongoing instructional process. Choosing to work in numerous clinical facilities, each of which has a narrowly specialized service, can lead to fragmentation of clinical experience, and to expenditure of excessive amounts of time on orientation to different settings. On the other hand, some change in clinical environment is desirable because it helps the student learn to adapt to different settings, and to avoid developing the attitude that "There *may* be different ways of doing things, but the way they are done in our hospital is best."

The question of faculty's moving from one ward to another, or from one agency to another, must also be considered. If a teacher changes the location of her clinical work frequently, she must not only spend considerable time orienting herself to different clinical settings but, because she is unfamiliar with the agency and its staff, she is hampered in developing an effective program of clinical instruction. Frequent changes of clinical environment can also make it difficult for the teacher to function with confidence and assurance, and to communicate these attitudes to her students. On the other hand, remaining indefinitely in one setting lessens her opportunity to develop ability to adapt to different clinical environments, and restricts the scope of clinical situations with which she has contact. The hazards of becoming involved in agency politics are also greater if the teacher remains in one setting over a prolonged period.

A choice must be made between continuity of contact between teachers and students, and continuity of the teacher's experience in a particular clinical environment. Continuity of experience in one setting allows the teacher greater opportunity to develop specialized clinical skills.

For example, the instructor who teaches only the care of premature infants has greater opportunity to develop clinical expertise in this area of nursing as well as orientation to the clinical environment. However, the span of time during which she works with each student is brief (thus minimizing continuity of instruction and opportunity for teacher and student to develop a constructive working relationship), and her teaching is limited to the care of one category of patients.

An effective orientation program does not focus primarily on the agency, with lessened emphasis on care of the individual patient and his family, and on the value of continuity of care from hospital to home. When preparing practitioners to function in a variety of settings, the emphasis should be on the patient's care, rather than on the particulars of how one agency functions. There is an important difference between developing constructive relationships at an agency through a useful orientation program, and narrowing the focus of orientation by concentrating on the agency's functions and facilities. When the latter approach is taken, the faculty may be prevented from recognizing other clinical resources which may provide more effectively for certain experiences than does the agency to which students are assigned; for example, a community well-baby clinic may provide more experience with normal infants than a hospital's pediatric clinic.

Interpreting the program to agency staff

Not only is it important for students and faculty to become acquainted with the agency, it is also essential for the agency staff to learn about the objectives of the particular clinical experience which students are having, and the kinds of clinical learning experiences which are sought by faculty in order to implement these objectives. Staff also need opportunity to discuss with the faculty how the clinical teaching will be conducted, and what role they are expected to have in the educational program.

The teacher in a new collegiate program is sometimes hampered in her efforts to develop creative ways of working with staff in planning students' clinical instruction. When these programs first develop in colleges and universities it is necessary to stress the responsibility of faculty for *all* aspects of teaching, both clinical and classroom. In the process, the ways that staff can contribute to students' education, without interfering with the faculty's fundamental responsibility for instruction, are sometimes overlooked. While there is no question that students' instruction is the responsibility of the faculty, there are many ways in which staff can contribute to students' learning. The staff nurse who is especially skillful in doing a new procedure can be asked to demonstrate it; the physician whose research is helping to refine cancer chemotherapy can be asked to talk with students about this aspect of therapy. Such experiences provide meaningful contact between staff and students, and a richer learning experience for students. As nurse educators and nursing service personnel more clearly define their roles in relation to students and patients, and as each group accepts the placement of nursing education in an educational institution, their energies can be more fully channeled into developing ways in which each group can contribute to the welfare of patients and students without interfering with their appropriate roles in education and service. Not only can agency personnel contribute to students' learning; the teacher can share with staff some of her particular clinical competence and interests. The rapid proliferation of knowledge makes such sharing a necessity in order to serve students and patients as effectively as possible.

In addition to discussing ways in which they can share clinical knowledge and skills, faculty and staff must discuss some everyday matters about the role each group is expected to fulfill while students are having clinical experience. For example, faculty should make it clear that a staff member who observes a student doing something immediately hazardous to a patient should interrupt it, but that in all other instances she should refer problems or questions concerning the students' performance to the teacher. Hence, one of the teacher's major tasks during the orientation period involves interpreting the purposes and methods of the educational program to staff of the agency. Sometimes there is considerable agreement between faculty and staff about the way students should be educated. This is not always the case, however, particularly when the program is different from the type with which staff have had experience.

Staff members sometimes express ambivalence about the program and about their beliefs concerning the education of nurses. For instance, nurses in a public health agency may state that they are glad to have an instructor present, because of the lack of time they have to spend teaching students. Their next words may convey, however, that they enjoy teaching, and miss the opportunities to teach students. A hospital staff nurse may say, "It is wonderful—the way students have so much instruction nowadays. But I know that I learned most when I was on alone at night. That's the way to really learn nursing."

Sometimes concern is expressed over irreconcilable alternatives. A nurse may say, "I wish I had had a chance to take liberal arts subjects but, on the other hand, these students do not get as much nursing practice as I had." Nostalgia for the good old days plays its part too, in interfering with understanding and acceptance of a new program. A physician may remark, "The registered nurses used to take care of my patients. Now the only time my patients see an R.N. is when the medicines are passed around. The more education nurses have, the less my patients see of them." One may be tempted to respond to such comments by implying that more and better education for nurses is the solution to all nursing problems, although one realizes that the gains from increased education will only gradually affect nursing practice, and that many other problems (such as discontinuity of practice and part-time practice) contribute to difficulties in providing nursing care. It is easy also to ignore the physician's observation about *who* is giving direct care to his patients by asserting, "But the purpose of our program is to prepare staff nurses" (while remembering perhaps that out of last year's class of 40 graduates only four are doing staff nursing in hospitals!) .

Discussion with physicians and agency staff can lead to sharing of ideas and recognition of mutual concern for the welfare of patients. For instance, one may acknowledge to the physician that baccalaureate graduates are often drawn away from the bedside to fill positions in teaching and supervision because there are not enough nurses prepared at the graduate level to assume these positions, and suggest that one way of alleviating the problem is to prepare more nurses (rather than fewer) at the baccalaureate and graduate levels.

Efforts to interpret the purpose of the program should be continued as long as students have experience at the agency, in order to orient new staff members, and to allow for gradual growth in collaboration between faculty and staff in developing the teaching program. Frequent informal discussions of the purposes of the program and of the effectiveness of various clinical experiences in achieving these purposes are just as important as the meetings which occur when clinical experiences are first sought in an agency, or in a particular department of an agency.

ESTABLISHING CONSTRUCTIVE PROFESSIONAL RELATIONSHIPS WITH SERVICE AGENCY PERSONNEL

Establishing constructive working relationships with the personnel of clinical agencies is an important facet of the teacher's work. The framework within which these relationships develop, and the relationships themselves, have been affected by the greater responsibility taken by faculty for clinical teaching, and by the increasing number of nurse faculty who are employed by colleges and universities. For example, a university faculty member who teaches students during their practice periods at a hospital has a different relationship with hospital staff from that of a head nurse who is responsible for clinical instruction and nursing service, or from that of the clinical instructor employed by a hospital to teach students in a diploma program.

The effectiveness of the plan of clinical instruction is affected greatly by the working relationships between faculty and agency staff. The teacher must consider which compromises are necessary in order to maintain a constructive working relationship, and which ones cannot be made, regardless of the strain which may be placed upon faculty-staff relationships. For instance, the teacher cannot permit a student to give an enema without a physician's order, even though the staff nurse states, "Since this is a long-term hospital, the doctors leave it up to us to decide which patients need enemas." However, if it is institutional policy to write the phrase "Morning care given" on every patient's chart, the teacher explains this policy to students and instructs them to carry it out. She should probably also discuss with students a different view of the purpose of nurses' notes—that of recording significant observations, rather than of listing usual aspects of care, such as bathing the patient. Many other situations are less clear-cut, in terms of whether the aims of the program are best served by modifying the usual policies governing student experience, or strictly adhering to them. For instance, how important is it to insist that students never leave the ward for supplies? In one hospital such insistence may be essential, because supplies are frequently not available on the ward. In such a setting, allowing students to go to central supply rooms, the pharmacy, and so on, could readily lead to their spending half of each laboratory period in such activities. Instead, the teacher should discuss the problem with nursing service personnel so that the needed supplies can be obtained in a way which does not interfere with students' learning.

The need for role clarification

In recent years emphasis has justifiably been placed upon the nurse's recognition and performance of her own role, rather than upon helping members of other professions carry out *their* roles; unless the nurse develops her own skills and recognizes her own contribution to patient care she has no basis on which to build a colleague relationship with others. Sometimes, however, in emphasizing the nurse's contribution to patient care, experiences which would enable the student to understand the roles of others are not given sufficient emphasis. It is just as undesirable for the student to lack understanding of the physician's role as it is for her to answer patients' questions with the stereotyped phrase, "Ask your doctor." Unless students have opportunities to learn about the roles of others who care for the patient, they acquire incorrect perceptions of their own role and also of the roles of others with whom they work. It is important, for instance, for the student to recognize that the surgeon's explanation and support are particularly meaningful to the patient because of his direct and often fateful therapeutic intervention, and because he possesses the fullest knowledge of the condition found during surgery, and of the patient's prognosis. This recognition in no way lessens the value of explanations and support which the nurse gives the patient, for example, by listening to his concerns about the surgery and helping him to understand and cope with changes in his body functioning which have resulted from surgery.

Another example of role clarification in fostering constructive working relationships with staff is the consideration which the teacher should give to her own participation in ward activities not directly related to students' assignments. Lending a hand in lifting a heavy patient, or handing a toddler her doll, are part of the nurse's role; to do these things in a setting where the focus is on patient care seems natural and appropriate. Offering this kind of assistance demonstrates to students how to make decisions concerning helping patients not assigned to them. Concentrating on their own patient assignments, and on their own role as learners, can be difficult for students, particularly when they see many patients who need care. Their desire to help can easily become a hazard unless they are advised against giving care for which they lack sufficient knowledge and skill, and against involving themselves in situations to which they are not sufficiently oriented. Observing how the teacher handles this problem helps students clarify how they should proceed. For example, if the teacher knows that the patient in an adjacent bed is allowed water, and he requests it, she gives it to him. However, she does not sweep past two staff nurses in the corridor in order to answer the call light of a patient whose care is their responsibility (thus possibly implying that she is attentive to patients while they are not). The teacher's attention is centered on care of patients assigned to students, and on teaching students. Helping other patients is done incidentally, and does not become the focus of her attention or detract from her teaching.

When she does not have a dual role in service and education, the teacher must scrupulously avoid giving opinions about nursing service problems—unless her assistance has been requested by the staff. To do so not only involves her in making decisions outside her sphere of responsibility, but it also invites staff to exceed the limits of *their* roles, and to involve themselves in making decisions about students' experiences.

Clearly defining and functioning within one's role are not the antithesis of sharing but, instead, provide a basis for sharing and collaboration. Possessiveness can interfere with sharing and with meaningful contacts which students should have with practitioners of the health professions. The student who works with a mentally retarded child over a period of time may be loathe to share her knowledge concerning the patient's care with staff. A public-health nurse may be reluctant to allow a student to participate in care of a family with whom the nurse has worked for many months. Possessiveness is demonstrated not only in relation to patients, but sometimes in relation to students as well. For instance, a teacher may discourage staff from contributing to students' learning, or she may not foster the beginning development of colleague relationships between students and staff; this is different from delineation by the teacher of learning experiences and her assumption of responsibility for providing these learning experiences.

Recognizing the differences in faculty and staff responsibility for patient care is important. When the teacher does not have a dual role, the staff are ultimately accountable for the patient's nursing care. Teachers' and students' responsibility for patient care, while definite at the time of their practice, does not exend over the span of the patient's illness. Leaving the agency after relatively brief laboratory periods serves to shield the faculty and students from some of the pressures and starkness of patient-care situations. Students and faculty must keep this fact in mind when they become critical of the way staff work with patients. The frustration of dealing with lack of needed equipment, for example, is easier to cope with for 16 hours a week than for 40. It is one thing to care for a dying patient for three hours; another, to be responsible for his care over an 8-hour period for several consecutive days.

The evaluation of student performance

Another aspect of the roles of faculty and agency staff which needs to be clarified—a major one meriting separate attention—pertains to evaluation of students' clinical performance. The teacher should make it known to staff that she is responsible for evaluating students' clinical work, but that she recognizes that she cannot observe every student all the time, and that staff also are observing students' practice. If staff make an observation which they wish to share with the teacher, they should feel free to do so, with the understanding that the teacher will use the information in a way which, in her judgment, best serves the student's learning. Unless staff are free to share their observations with the teacher, important observations which could be used to help the student may be lost, or recognition of problems in students' performance may be delayed. Mounting tension between faculty and staff can occur because failure to discuss questions and problems daily can lead to misunderstandings about what each group expects of students.

Frankness in discussing problems

Over the years, the teacher's accountability for the care students give to patients has changed in many schools. In situations where formerly the instructor was responsible only for classroom teaching and for conducting clinical conferences, the head nurse and supervisor were responsible for such matters as noting that the students gave medicines accurately and on time, and that side rails were in place. As the teacher has become more involved with direct clinical instruction and supervision, however, she has also become more accountable for the manner in which students carry out patient care.

The more direct the teacher and staff nurses can be in discussing their concerns and disagreements about care rendered by students, or policies and practices on the ward which affect students' practice, the better. Faculty and staff should discuss problems together daily and work out solutions. It should not be necessary to involve administration in these details. To do so indicates that faculty and staff are not confronting issues together but are avoiding such discussions by referring matters to their respective administrators. Assistance should be sought from administrators when the problem is not within the province of faculty-staff functioning, or when faculty and staff cannot arrive at a satisfactory solution.

Directness between faculty and staff necessitates a mutual understanding that tale-bearing will not be resorted to as a way of dealing with problems which may arise in their working relationship. Both faculty and staff have opportunity to observe each other's functioning at closer range than is ordinarily true of the dean's observation of her faculty's clinical work, or the director of nurses' observation of her staff's performance. This makes it inevitable that lapses in performance come to the attention of each group. The head nurse's observation that a student has not recorded intake and output, or the instructor's observation that an aide removed a footboard which the student had carefully provided for a patient, should be dealt with by the head nurse and teacher. Faculty and staff can best work together if they know that each will go first to the other with such problems, and that others will be involved in the discussion only if problems cannot be handled in this fashion.

Communication between faculty and staff about the specific aspects of care which students will carry out is essential. It is important for both faculty and nursing staff to recognize that students care for patients for limited periods of time, and that during these periods, only selected aspects of care are undertaken. The more inexperienced the student, the fewer are the facets of care which she can carry out. Assignment of some aspects of a patient's care to a student should not mean that the patient is "crossed off the staff's list," but instead, that staff are aware of the care the student is giving, are ready to resume these aspects of care when the student leaves, and continue to carry the responsibilities for the patient's care which the student cannot assume, either because of her inexperience, or because of the limited time she spends at the agency.

Opportunities for students to work with agency staff must be deliberately fostered by the teacher in order to help the student learn to coordinate her role with that of others who care for the patient. Working with staff is especially important for students in collegiate nursing programs because, due to the limited time they spend in clinical practice, they often have fewer opportunities for such contacts than students in diploma programs.

Participation of agency personnel in
planning learning experiences

It is important to distinguish between the role of faculty in making decisions concerning educational experiences for students in their courses, and the responsibility of the profession as a whole to develop long-range plans for the education of future practitioners. Education of students is a legitimate concern of all nurses because it affects the future of the entire profession. Dialogue between nursing educators and nursing service personnel should be fostered and, through it, decisions reached concerning the future education of nurses. However, during daily clinical experience the teacher must make decisions which are solely her responsibility. Whether the student will care for two patients or six, or whether the newly admitted patient with bleeding esophageal varices should be assigned to a student, rests entirely with the teacher. Her approach in such situations should be definite and firm, and should avoid engaging the staff in debate. (This does not imply that the teacher should not ask staff for suggestions about suitable patients, but that she must assume the responsibility for deciding which learning experiences are suitable.) Arguments between faculty and staff over issues which have already been decided concerning the present student experience are fruitless. For example, the staff may believe that students should spend four days a week on the ward. If an agreement has already been reached between a college and an agency for two days' clinical experience a week, daily discussion between teacher and staff about the lack of time students have for clinical practice is pointless. Instead, their efforts should be directed toward planning for most effective use of the two days available for students' clinical work, and making recommendations concerning the amount of practice desirable for future groups of students.

Development of satisfying working relationships takes time and effort on the part of faculty, staff and students. Such relationships flow from many things—informal talks between head nurse and teacher over a cup of coffee, consideration of each group's comfort and responsibilities, and the sharing of interest and concern about the welfare of patients and students. Perhaps most important is the development of mutual respect between faculty and staff, which is based upon recognition of the differences in their roles, and the belief that each group has a useful contribution to make within the profession of nursing.

SUGGESTED READING

Fonseca, Jeanne D. "Faculty Planning for Field Practice," *Nursing Outlook, 13*:56, October, 1965.

Montag, Mildred L. "Nurse Faculty in Associate Degree Programs," *Nursing Outlook, 12*:40, July, 1964.

Redman, Barbara K. "Conflicts in Clinical Teaching in Nursing," *Nursing Forum, 4*:48, No. 2, 1965.

Scher, Maryonda, and Nehren, Jeanette. "A Student Experience That Taught Faculty and Staff," *Nursing Outlook, 14*:26, July, 1966.

Index